T0186892

# WATER STRESS: SOME SYMPTOMS AND CAUSES

העניים והאביונים מבקשים מים ואין
לשונם בצמא נשתה
אני יהוה אענם


لو كنت تعلمين عطية الله
و من هو الذي يقول لك أعطيني لأشرب
لطلبت أنت منه فأعطاك ماء حيا


ΕΑΝ ΤΙΣ ΔΙΨΑ ΕΡΧΕΣΘΩ ΠΡΟΣ ΜΕ ΚΑΙ ΠΙΝΕΤΩ
Ο ΠΙΣΤΕΥΩΝ ΕΙΣ ΕΜΕ ΚΑΘΩΣ ΕΙΠΕΝ Η ΓΡΑΘΗ
ΠΟΤΑΜΟΙ ΕΚ ΤΗΣ ΚΟΙΛΙΑΣ ΑΥΤΟΥ ΡΕΥΣΟΥΣΙΝ
ΥΔΑΤΟΣ ΖΩΝΤΟΣ

# Water Stress: Some Symptoms and Causes

## A case study of Ta'iz, Yemen

CHRIS D. HANDLEY
*School of Oriental and African Studies*
*University of London, UK*

**Ashgate**

Aldershot • Burlington USA • Singapore • Sydney

Published by
Ashgate Publishing Limited
Gower House
Croft Road
Aldershot
Hampshire GU11 3HR
England

Ashgate Publishing Company
131 Main Street
Burlington, VT 05401-5600 USA

Ashgate website: http://www.ashgate.com

**British Library Cataloguing in Publication Data**
Handley, Chris D.
    Water Stress : some symptoms and causes : a case study of
    Ta'iz, Yemen. - (SOAS studies in development geography)
    1. Water-supply - Yemen - Ta'iz 2. Water-resources
    development - Environmental aspects - Yemen - Ta'iz 3. Water
    conservation - Yemen - Ta'iz
    I. Title
    333.9'1'09533

**Library of Congress Control Number:** 00-111836

ISBN 0 7546 1524 3

Printed and bound by Athenaeum Press, Ltd.,
Gateshead, Tyne & Wear.

# Contents

v

# List of Figures

# List of Tables

# Preface

As the World population increases in number, per capita environmental resources, such as water, decline. Simultaneously, as people aspire to more comfortable lifestyles, those resources are returned to the environment in a more degraded state than when they were taken, resulting in pollution. The Yemeni city of Ta'iz (population 400,000) has experienced both these processes to a marked and essentially unchecked degree.

This study suggests that in order to develop water resources sustainably in areas facing water shortage, an understanding of the factors leading to scarcity require an integrated, interdisciplinary and holistic approach. This hypothesis has been tested in the context of the water shortage crisis in Ta'iz that peaked in summer 1995. The crisis resulted from the demise of the main aquifer supplying the city. Water supply declined to deliveries of less than once every forty days. Per capita consumption and the quality of the water were below WHO recommended minima.

The impact of water use and misuse on the environment of Ta'iz and hence on those who share it has been extreme, perhaps too extreme to be considered relevant to most people. However, as the causes of water scarcity touch other parts of the World, the lessons learnt in Ta'iz may serve as an early warning and, it is hoped, contribute to averting them.

Chris D. Handley
Ta'iz
2000

# Acknowledgements

I would like to acknowledge Dr Mohammed Lutf Al Eryani for making the research in Yemen possible and Abd Ar Rahman Al Eryani for giving insights into the political makeup of the field area. Abd As Salam Abd Al Alim Hasan provided valuable exposure to indigenous legal and institutional frameworks, Abdullah Saif provided hydrological data, and Drs Gazi As Saqqaf and Khalid Riaz agricultural economics data. Digby Davies, Christopher Ward, Cecile de Rouville and Drs Aslam Chaudhry, Marcus Moench, Alex McPhail, Tony Zagni and Peter Mawson are gratefully acknowledged for stimulating discussion and assistance in understanding socio-economic issues related to the study. Drs Richard Morris, Malcolm Lonsdale, Barry Lakeman and John and Andrew Mitchell, and Roy Avis, Gavin Thomas and Debbie Lakeman generously gave of their time in discussions, statistical and meteorological advice, support and logistics. Office facilities were made available by Margaret Shaw and the Kerith Centre. Jane Dottridge (UCL) patiently guided chapter two, and critically reviewed the book with Dr Linden Vincent. Dr Hiro Yoshida and Yasser Muheildeen assisted with the satellite imagery and the preparation of the map figures, I would especially like to thank Gerhard Lichtenthaler for his continued encouragement, my wife for her patience and proof-reading and last, and perhaps most, Professor Tony Allan for his supervision, provision of facilities, encouragement and support.

I am also grateful to UNDP, the World Bank, GTZ and Dar El-Yemen for permitting me to use data collected whilst on contract to them and to the staff of the Geography Department, SOAS for helping facilitate the task.

# List of Abbreviations

| | |
|---|---|
| AREA | Agricultural Research and Extension Authority |
| CACB | Cooperative and Agricultural Credit Bank |
| CES | Consulting Engineers Salzgitter GmbH |
| CPR | Common Pool Resources |
| CROPWAT | Computer program for irrigation planning and management |
| CSO | Central Statistical Office (Ministry of Planning) |
| DISWC | Department of Irrigation and Soil and Water Conservation |
| EC | Electrical Conductivity |
| EEC | European Economic Community |
| EKC | Environmental Kuznetz Curve |
| FAO | Food and Agricultural Organisation |
| GAREWS | General Authority of Rural Electricity and Water Supply |
| GDP | Gross Domestic Product |
| GPS | Geographical Positioning System (Hardware) |
| GTZ | Deutsche Gesellschaft fur Technische Zusammenarbeit |
| GWV | Ground Water Vistas (Software) |
| HP | Horse Power |
| HWC | High Water Council (of Yemen Arab Republic) |
| IDAS | Innovation Development in the Agricultural Sector |
| LC | Local Council |
| LCCD | Local Councils for Cooperative Development |
| LDA | Local Development Association |
| LWCP | Land and Water Conservation Project |
| MENA | Middle East and North Africa (World Bank Region) |
| MEW | Ministry of Electricity and Water |
| NGO | Non-Government Organisation |
| NIE | New Institutional Economics |
| NWRA | National Water Resources Authority |
| NWSA | National Water and Sanitation Authority |
| PE | Potential Evaporation |
| PPP | Polluter Pays Principle |
| PRA | Participatory Rural Appraisal |
| PSP | Private Sector Participation |

| | |
|---|---|
| RASM | Readily Available Soil Moisture |
| RO | Reverse Osmosis |
| RRA | Rapid Rural Appraisal |
| SCS | US Soil Conservation Service |
| SOAS | School of Oriental and African Studies |
| SURDU | Southern Uplands Rural Development Unit |
| TOR | Terms of Reference |
| TS-HWC | Technical Secretariat of the High Water Council |
| TSWSSSR | Technical Secretariat for Water Supply and Sanitation Sector Reform |
| TWSSP | Ta'iz Water Supply and Sanitation Project |
| UCL | Universtiy College London |
| UNDDSMS | United Nations Department for Development Support and Management Services |
| UNDP | United Nations Development Programme |
| WDM | Water Demand Management |
| WHO | World Health Organisation |
| WIER | Water is an economic resource |
| WINER | Water is not an economic resource |
| WRM | Water Resources Management |
| WTP | Willingness-to-Pay |
| YR | Yemeni Riyal (Currency) |

1. During the main period of field work (1995 to 1998) the exchange rate for the Yemeni Riyal varied from 120 to 140 YR = $1US. 130 YR has been used.

2. Although the literature prefers North – South terminology to West – East, for a meaning of Developed World – Developing World, West – East is used here. "North – South" has been reserved to mean pre-unification states of Yemen (pre 1990) or after that date, and also generally, to the peoples North and South of Yarim.

# 1 Introduction

A meeting was held in Ta'iz, Yemen, in observance of the World Day for Water, 1998. The meeting was attended by representatives from major organisations associated with foreign aid and development, local community and local and central government representatives, NGO's involved in environmental protection, and, in the name of public awareness, several hundred noisy school children. The meeting was hosted by the relatively newly established national body responsible for water resources. After the meeting, a group of smartly dressed people representing most of these organisations went on a brief field trip in a rather new and luxurious Layla 'alawi (latest Toyota Landcruiser status symbol, named after a Yemeni politician's daughter who acquired one). The chosen destination was a 400m deep borehole in Habir being tested to determine its safe yield with a view to connecting it to the main pipeline supplying the city of Ta'iz 25 km to the South. After a brief conversation with the engineer responsible for the test, a nearby shallow well dug into the wadi gravels was visited, from which its owner happened to be irrigating a field of tomatoes. Standing in his farm-soiled local dress he informed the group with great conviction regarding the detrimental effect the pumping test was having on his water supply. The group then returned to the Governor's lunch awaiting them in the city of Ta'iz, the 400,000 inhabitants of which, received poor quality water once every 3 weeks from the public utility.

On the way from field to city the plight of the (assumedly) poor farmer was discussed by the eminent experts. Depending on the background of the speaker, discussion topics ranged around how soon the well could be on-line, the need to compensate locals for derogation, the role of appropriate stakeholder representative local institutions, the need to understand the locals' perceptions and social water uses, the application of Islamic law to the situation and many others. However, the discussion seemed to miss the reality. Actually, monitoring of the farmer's well indicated no effect after extended pumping from the test well in terms of water level or hydrochemistry. He was simply being opportunistic.

1

Although the visitors had failed to determine this information, vital as it was to their grand strategising, it was not the central issue. More importantly, the visiting actors managed to reach the wrong conclusion by following the scripts and narratives their disciplines had taught them, resulting in their missing the "big picture" of the links the common subject of water must form between those disciplines if water resources management is to be effective. The incident described above not only contains in microcosm some of the multi-faceted aspects of water resources management involved in the area of Ta'iz, but also, and, more importantly, the need for the managers to have a multi-faceted grasp of their task.

## Purpose and Scope of the Study

A central hypothesis to this study is that an integrated, interdisciplinary approach to water use is needed if determinants of allocation are to be understood and sustainable measures introduced. The water stressed situation which has evolved in Ta'iz between 1965 and 1995 provides an example with which to test that hypothesis. The Ta'iz data are also used to examine critically:

1. the relevance of demand management,
2. the role of adaptive capacity, particularly social,
3. the relative importance of economic and political factors,
4. the contribution of plural legal and institutional frameworks,
5. the significance of virtual water and population growth, and
6. the potential for sustainable development.

in the allocation of water in the context of severe water stress in a Southern state.

The water resources of a specific area (Wadi Al Hayma, 16 km$^2$) are evaluated within the context of the record of abstraction for domestic use and, using satellite imagery and water balance modelling, for agricultural use as well. The environmental cost of the depletion of the main aquifer in Al Hayma is evaluated in terms of projected lost agricultural production that would be incurred in enabling aquifer recovery. The returns to water from industrial use are analysed numerically and contrasted with those from agricultural use for the wider (930 km$^2$) Upper Wadi Rasyan catchment. [The locations and areas of Wadi Al Hayma and Upper Wadi Rasyan included in this study are indicated in Figs. 3.1 and 3.2.] The environmental impact of industrial and urban domestic water use is also qualitatively assessed. The politicisation of water allocation between agricultural and urban domestic users and the potential for conflict are critically examined. The dependence on virtual water to meet the food needs of the increasing population is quantified. The plural institutional and legal frameworks are appraised from the perspective of whether they contribute to providing or preventing equitable water allocation. The efficiency of the water markets and whether their provision is equitable are critically examined and the volume of water transacted on the market, and its quality and price are evaluated quantitatively. The capacity to adapt to water shortage across the spectrum of social scale from individual households and businesses to the government is portrayed statistically and through an examination of specific allocative issues respectively. A review of past and present economic development efforts permits a critical assessment of sustainable development models and the potential for environmental protection and equity provision.

## Study Outline

No doubt the literature would be more replete with the useful lessons learnt from failures if professionals were not so economically and politically insecure. In many fields the study of failures offers the greatest opportunity for understanding processes. Ta'iz is an example of failed water resources management. The author of this study was privileged to be able to observe the consequences of that failure. The fieldwork spanned the period 1995 to the end of 1998. Although the "water crisis" peaked in the summer of 1995, it continues to this day and yet higher peaks may lie ahead. Some causes of the crisis are traced from 1960s roots, although an earlier beginning is explored to explain some of the deeper causes.

Many factors have contributed to the water problems of Ta'iz and the "knock-on" effects of those problems have been numerous and diverse. An analysis of the problems, their causes and effects demands an integrated, holistic examination of their linkages. The locals and the passing observer view the "whole picture" as it faces them. Specialists from the disciplines of hydrogeology, agriculture, development, economics, politics, environmental and any other field involved notice the colour of their discipline that contributes to the painting. The policy maker also wants to see the whole "water colour" and needs to understand the linkages between the various aspects of it that would traditionally belong in different academic pigeonholes. The study attempts to examine the nature, extent and origins of the shortage. Attempting to integrate the different fields which contribute to the Ta'iz water crisis provides a unique opportunity to examine the linkages between many diverse physical and human aspects of water resources management in the context of extreme water shortage.

The physical water environment that Ta'iz impacts, and affects Ta'iz, is investigated in terms of water availability and quality in chapter two. The development of the most important aquifer to Ta'iz is examined in detail and a reconstruction of the causes of its demise is attempted. The social response to water availability and use is considered against the underlying economic realities in chapter three. These two chapters provide physical and socio-economic "maps" of the shortage. Because different methodologies were used to create the "maps" of the shortage, each is described in its relevant section, and there is no single "methodology" section. In chapter four, the concepts and principles discussed in the literature are used as filters for looking at the Ta'iz "maps". The "maps" are also used as a means of testing and revising the concepts and principles. Chapter four is structured on the basis of the proposed causative sequence described below, but in the direction of symptom to cause rather than vice versa. The Ta'iz experience is considered in the context of the apparent contradiction of sustainable development. Chapter five concludes the study by considering the appropriateness of the integrated approach and asking whether the lessons learnt from Ta'iz could be of value to other population centres facing similar crises.

## Initial Concepts

The lateral and vertical distribution of people vis-à-vis water provides a means of attempting the contradictory; the compartmentalising of something which is inherently holistic, that is, water.

### The Lateral Distribution of Water and Humans: Urban – Rural Distinctions

At the most basic level, *homo sapiens* interacts with the environment extracting needs and returning waste (Table 1.1).

**Table 1.1 Man interacts with the environment**

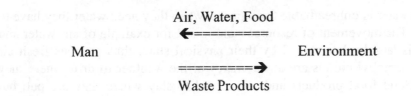

Human communities, seeking to enhance their comfort, shelter, convenience and security, tend to concentrate in settlements. These settlements must look to ever increasingly technological means to provide man's needs (and wants). The location supplying the basic needs typically lies outside the settlement, and the "footprint" of supply and environmental impact spreads as the settlement grows. In terms of water this is probably better termed a maldistribution of people than of rainfall (Turton, 1999a) and results in a rural - urban (and even West - East) distinction (Table 1.2):

**Table 1.2  Urban – rural distinctions**

| Urban | Rural |
|---|---|
| Resource Demand Centre | Resource Supply Footprint |
| Waste Supply Centre | Waste Disposal Area* |
| Industrial Activity | Agricultural Activity |
| Urban Domestic Supply | Rural Domestic Supply |
| Increasing Livelihood Provision | Declining Livelihood Provision |

Result:      Population Pull  ⬅==== Population Push    = Urbanisation

* The urban area can also be one of waste disposal, but the "footprint" of environmental impact by waste may spread beyond the area occupied by the urban waste producers (Serageldin, 1994; 5).

Lundqvist (1998; Table 1) makes a similar urban-rural distinction using the terminology of industrial-mechanical and biological-landscape, differentiating waste disposal (externalities) as being diffuse and concentrated respectively.

*The Vertical Distribution of Water and Humans*

Since water is unbreathable for humans, when they need water they have to lift it. The movement of resources and waste, for example of air, water and food is largely determined by their physical state, thus gaseous fresh air and Chernobyl clouds are at the mercy of the weather to drift where they will. Solid food products and waste tend to stay where they are put, but liquid water supplies and polluted waste water flow down-hill under gravity unless impounded by technologically adequate means.

The primary dynamic acting on water is the physical one of gravity. However, 'water flows up-hill to money and power' (Reisner,1986). In attempting to reverse gravity-flow, human demand (an "anthropocentric" dynamic, or "egocentric" dynamic when want exceeds need) is met by technological innovation. As well as a preference to be upwind, upstream and as far as possible from a landfill site, Yemeni settlements, unlike temperate climate settlements, tend to be located in mountainous areas and are often on the mountain ridges. This may be for defence reasons and also to keep away from the dangerous flash floods of the wadis. The result is the need for a lot of water to be lifted. (Allan, 1994a notes a similar population / water elevation separation on a larger scale in the Jordan catchment). The rainfall distribution also results in communities on the mountains being nearer to the ultimate source, that is the rain, and enjoying the economic advantages associated with being upstream users rather than downstream users (Varisco, 1983). The poor, who often seem to be located at the downstream end and are missed out by the anthropocentric dynamic, are sometimes provided for by religious or areligious equity-driven do-gooding. This third, least evident, and therefore weakest dynamic perhaps could be called "good" or "theocentric".

*Dissecting the Holistic for the Purpose of Analysis*

For water resources management to contribute constructively to the establishment of an economy developing in a sustainable manner the unenviable task of policy formulation and decision making must weigh a host of variables, some of which are inherently in tension. The need for holism in determining those weights is self-evident, but the variables still have to be isolated to be weighed. As a starting point primary aims must be distinguished from secondary and ends from means. The following table summarises some basic aims in providing water:

**Table 1.3 Aims of water provision**

| Adequate Provision | Environmentally Sustainable Provision | Equitable Provision |
|---|---|---|

In providing water there are two ways of making it go further. One is to make it produce more of the same (productive efficiency) and the other to produce something else of greater value (allocative efficiency). "The same" is typically defined as within the same broad sector, such as agriculture, rather than the same item, potatoes, for example.

**Table 1.4 Methods / means of efficient water use**

| Productive Efficiency | Allocative Efficiency |
|---|---|

Various mechanisms contribute to both allocative and productive efficiency, though engineering ones tend to be directed more at the latter and fall in the category of supply management measures (that is, which increase the supply). The dynamics acting on water provision are shown in Table 1.5.

**Table 1.5 Dynamics of water provision**

| Determining Dynamic | Gravity | Money | Power vs. Equity Coercion vs. Co-operation |
|---|---|---|---|
| Context | Physical | Economic | Cultural / Socio-Political |
| Mechanisms | Engineering | Signals | Legal Framework – "Rules" |
| Enablers | Projects | Markets | Institutions – "Players" |

These forces operate within, and help to shape, a context which is not static and in which various allocative trends or processes are taking place. The removal of water subsidies, creation of water markets and establishment of reallocative water transfers are economic measures which can help reduce demand, but which all require political facilitation. Political enabling, in turn, is conducted by institutions directed and supported by a legal framework. This sequence is depicted as:

Political => Legal => Institutional => Economic =><= Social =><= Physical

For example, the outer layer of this sequence, the physical environment, is perhaps the easiest to observe and measure. Abstracting resources from it and returning waste to it is the sphere of human activity, which is in turn influenced by underlying economic realities. In a democratised society, such as the UK, the economic realities of water transactions, for instance the magnitude of one's water bill, are meant to reflect rates set by the relevant institution, for instance the water authority or water company. Thus the institutional layer underlies the economic. That institution has a legal mandate, for instance from OFWAT, which is in turn established by a legal framework (the next layer), which itself came into existence through the deliberations of an elected political body (the core). The extent to which this sequence of causation allocates water, and thereby the benefits of water use, equitably or for the perpetuation of power asymmetries is a key issue.

The sequence is used here merely to provide a means of examining a whole. Some of these categories are rather loose, may overlap to an extent, and may be short-circuited. In some instances the occurrence of feedback suggests the relationships between them are interactive (such as those in power responding to the needs of the electorate). For the purpose of this discussion "physical" refers to the quantity and quality of water, that is, the water environment, with which communities interact. Together with economic signals, or their absence, these environmental constraints largely determine the hydraulic activity limits of those communities (for instance limiting the extent to which the desert can be made to bloom).

When compared with other rather unidirectionally changing trends and processes that typify the East, some of which are listed in Table 1.6, the mechanisms and contexts in which the compartments of the sequence operate are much more static.

The trends shown in Table 1.6 have been evident in Yemen, particularly during the past thirty years. During this period economic development has entered a water resources management arena in which the various contexts, mechanisms and trends mentioned above are already operating. As a consequence, economic development, which seeks to modify the status quo, is itself modified by the limitations of that arena.

**Table 1.6 Consistent eastern trends and processes**

**Trends / Processes**
Population Increase
Urbanisation
Technology / Communication

## The Global Position

*Global Water Statistics Indicate the Size of the Problem*

Allan (1994a; 4) points out the huge differences in annual human water needs: for drinking, $1m^3$; for domestic purposes, $100m^3$; and in food, $1000m^3$. To meet those needs plus industrial demand, Ohlsson (1995; 5) estimates that around one third of the world's 12-14 thousand $km^3/yr$ of non-soil, renewable freshwater resources are withdrawn. He breaks down these withdrawals as 69% to agriculture, industry 21%, and municipal 6%, however, sector consumption is 89% to agriculture, industry 5% and municipal 2%. The huge consumption by agriculture is exacerbated by irrigation. However, 36% of agricultural yields come from the 16% of farmland which is irrigated.

There are really only two ways of "losing" water, either through evaporation (from which state it will rain on someone else) or through contamination. Seepage of aquifers to the sea and the flow of rivers to the sea are the equivalent of loss through contamination. The assessment of soil-water in global water statistics is referred to by Allan (2000). Its under-utilisation can only occur if evaporation is taking place instead of evapotranspiration. Fallow ground therefore is a source of water loss to an area. If the soil cannot retain enough water to support the crop for the full season and there is no potential to supplement the water supply then it is difficult to see how the soil water could be utilised. Some water will inevitably be "lost" to evaporation from the soil. That which is "lost" through seepage to the aquifer could be reclaimed but only with energy input for lift.

In 1991, Vincent (1991; 197) stated that only 1.3Bn of the world population had clean water and 700m sanitation, noting that coverage was especially poor in the Middle East. Serageldin (1994; 3) suggested that 1Bn people lacked an adequate water supply and that 1.7Bn were without sanitation. Lundqvist (1998) states that 1Bn are without access to safe water and 800m are without a secure food supply. Although many global water statistics are bandied about, some of which are contradictory, whilst others do not compare like with like, the overall picture of water shortage and an increasing drain on the resource base is consistent. The consequences of shortages are literally a matter of life and death. Within the MENA area (World Bank designation for the Middle East and North Africa) Berkoff (1994; 14) considers that the provision of uncontaminated water supplies could reduce death in rural and urban areas by 30%, and another 20% if sanitation were to be provided. Most of the deaths are amongst the 0-14 year age range who account for 43% of the area's population.

Population growth is the single most important factor affecting the above estimates. Ohlsson (1995; 28) estimates a world population of over 10Bn by 2050, 96% of whom will be in the South. Allan (1994b; 66) bemoans the emphasis by regional leaders on the water supply deficiency whilst de-emphasising the demographic explosion that is its cause. However, the global picture is not all gloom. Despite a population increase of 23% in the developing world between 1980 and 1990, access to safe water increased from 77 to 82% and from 30 to 63% in urban and rural areas respectively and sanitation coverage increased from 69 to 72% and from 37 to 49% in urban and rural areas respectively (World Bank, 1993; 36).

*Some Attempt to Solve the Problem*

The reason for such increases in safe water and sanitation provision in the 80s was partly due to the efforts of the global development community in response to the "Water Decade". The declaration that the 1980s should be designated as the International Water Supply and Sanitation Decade occurred at the UN Water Conference in Mar Del Plata, Argentina in 1977. The conference was described as 'the first of its kind ever held in the area of water' which 'sensitised the world community on the importance of water for development' (Thanh and Biswas, 1990; xiii). This was followed by the UNDP Global Consultation on Safe Water and Sanitation for the 1990s in New Delhi at which participating musketeers aimed at 'some...' safe water 'for all rather than all for some...' by the year 2000. Four parts of the conference statement (quoted in Vincent, 1991; 212) reflect the state of play in water debates:

1.     'Protection of the environment and safeguarding of health through the integrated management of water resources and liquid and solid wastes.'

2.     'Institutional reforms promoting an integrated approach and including changes in procedures, attitudes and behaviour, and the full participation of women at all levels in sector institutions.'
3.     'Community management of services, backed by measures to strengthen local institutions in implementing and sustaining water and sanitation programmes.'
4.     'Sound financial practices, achieved through better management of existing assets and widespread use of appropriate technologies.'

Vincent is particularly critical that the statement suffers from 'confusing and contradictory rhetoric' and anticipated that it would have no impact. She cites the 'mixed bag' of item 2 measures as 'confusing' and item 4 as 'contradictory and controversial'. However, the points did focus on the main water issues of our times, or at least on those of the changing agenda of the aid fraternity. According to Serageldin (1994; 1), the previous "old agenda" had been about providing household services, whilst this "new agenda" emphasises environmentally sustainable development.

    Other international conferences have also influenced the "new agenda". In 1991, the UNDP held the Delft conference entitled A Strategy for Water Sector Capacity Building which dealt primarily with the need for reform in water institutions. The International Conference on Water and the Environment held by the UN in Dublin in 1992 felt that 'it is vital to recognise first the basic right of all human beings to clean water and sanitation at an affordable price...' (quoted in Lundqvist, 1998; 9) and stated the new agenda rather more clearly in what are widely referred to as the Dublin principles:

1.     Fresh water is a finite and vulnerable resource, essential to sustain life, development and the environment.
2.     Water development and management should be based on a participatory approach, involving users, planners and policy-makers at all levels.
3.     Women play a central part in the provision, management and safeguarding of water.
4.     Water has an economic value in all its competing uses and should be recognised as an economic good.

The water resources policy measures of Delft and Dublin were endorsed by world leaders at 1992 UNCED in Rio de Janeiro and were included in the Earth Summit of 1993 (Le Moigne et al, 1994; 3). McSweeney (1998; 10), summarises Dublin as identifying that water is 'to be treated as a scarce valuable resource...' and that the conference marked the 'introduction of demand management mechanisms...'. He also describes Rio as 'the most comprehensive international environmental and social declaration to date...' and as having 'equity at its heart...', by which he means "intergenerational equity": that is, not dumping the consequences of our environmental misuse on future generations. Although Rio diluted Dublin it did identify the fundamental dilemma in the regulation of competitive water use: conservation of water as a prerequisite for development versus development threatening water availability (Du Bois, 1992; 1).

Briscoe (1994) succinctly describes the "new agenda" of the international conferences as 'no more than the application, to water, of the great ideas of our time, namely democracy and the market...' (quoted in Davies and Sahooly, 1996; 9). However, whether the slippery concept of equity at the heart of Rio will become the 'greatest good to the greatest number of people...' (McSweeney, 1998) is questioned in Orwellian style by Allan (1994a; 3) when he asks 'equity for whom...' and 'on what scale?'

# 2 Autopsy of an Aquifer

## Introduction

The Ta'iz study area is an extreme case of environmental degradation and provides a unique opportunity with which to examine it's causes. The demise of the water supply of Ta'iz has centred on two main water resource problems. Firstly, the resource base has been declining, and secondly, that resource has become increasingly polluted. This chapter investigates both aspects, but the major part examines the declining groundwater levels in the main aquifer serving the city, that of Al Hayma (Figure 2.1) in order to:

1.  provide an understanding of the natural water movement processes,
2.  assess the relative significance of human water use activities,
3.  determine the physical causes of the crisis,
4.  define the sustainable yield of the aquifer, and
5.  predict how long it would take the aquifer to return to its initial state if current minimal abstractions for the city were to continue.

The latter point permits evaluation of the cost of reversing environmental degradation in chapter 3. The problem of declining surface and groundwater quality is described and its causes identified, and the chapter concludes with a summary of the environmental impact of water use and abuse in the Upper Wadi Rasyan catchment (area defined in Figure 2.2).

### Groundwater Development: Historical Background

Since the 1960s, the water supply of Ta'iz has been derived from groundwater sources located successively further north of the city. The earliest wellfields were developed close to the city (Hawban / Hawgala, Figure 2.1) but were of low quality water. Water quality has deteriorated further because the wellfields are located downstream of the city and receive effluent from it. Development then shifted to the Al Hayma valley

12 km to the north. As this source was exhausted, the conveyor was extended further north to the Habir area where the Tawilah Sandstone was hoped to provide some relief to the water shortage. Exploration further afield continues to today. A summary of these events follows (Box 1):

---

**Box 1: History of Hydraulic Events in Ta'iz**

Early 1960s: Hawban and Hawgala fields developed and Kennedy scheme developed to supply Ta'iz (low quality water) by USAID.

1967 USAID leave, Kennedy scheme starts to deteriorate.

1975 J M Montgomery recommend Al Hayma–Miqbaba wellfields.

1976-81 Investigation of Al Hayma wellfield (Miqbaba – As Sahlah) recommends utilisation of all groundwater in Al Hayma including the complete cessation of irrigated agriculture. Tipton and Kalmbach (1979) recommend compensation of $10M to farmers (which was never paid). Awareness of the impossibility of simultaneous use by agriculture and Ta'iz city.

1982-3 Commissioning of Al Hayma wellfield.

1987: Al Hayma wellfield running dry. Emergency measures include reconnection of low quality Hawban Hawgala fields and emergency drilling of six more wells in Al Hayma.

1989: Attention turned to Tawilah Sandstone in Habir, immediately north of Al Hayma. Negotiations for exploratory drilling begun with shayxs of Habir / Dhi Sufal. Some exploration wells drilled.

1995-6: Water supply in Ta'iz reaches once per 40 days, EC >1500µS/cm. Emergency drilling in Ta'iz city by instruction of the Governor. Negotiations with shayxs of Habir / Dhi Sufal for six wells to be connected to an extended Hayma – Ta'iz conveyor. Three connected (from West Habir).

1997-Present (1999): Declining yields of three West Habir wells and Ta'iz city wells. Exploration by NWRA of outlying areas.

---

*Sources*: Leggette et al (1977 and 1980), Montgomery (1975), Adel (1986), Dubay (1989, 1993 & 1996), Tipton and Kalmbach (1979), CES (1997).

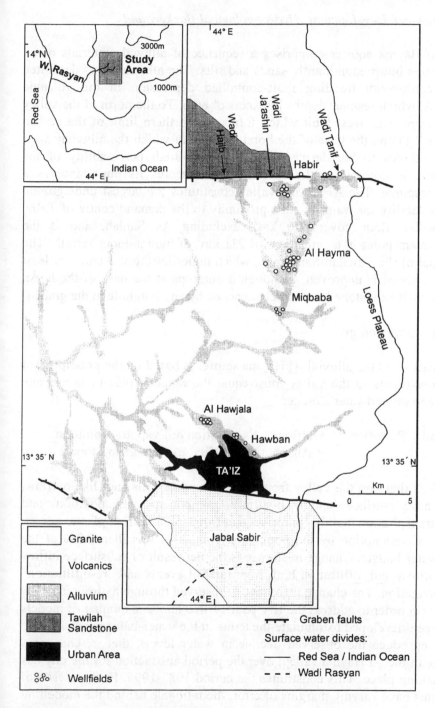

**Figure 2.1      Location map**

## Groundwater Development: Hydrogeological Background

The Al Hayma aquifer comprises a sequence of alluvial materials of all grain sizes but predominantly sands and silts. The alluvium was deposited in a north-south trending fault-controlled depression in the volcanic bedrock, which reaches depths in excess of 80m. To the north of the valley lies a major east-west fault which forms the northern limit of the graben (Figure 2.1) and the base of the horst slope from which the alluvium was, and continues to be, derived. The relatively high permeability of the alluvium, in contrast to the volcanics, underscores its importance as a water resource. The Al Hayma valley constitutes the largest underground storage facility for water in such proximity to the demand centre of Ta'iz. The valley floor covers 16 km$^2$, excluding As Sahlah, and is the downstream point in a catchment of 231 km$^2$ of mountainous terrain. The potential of the Tawilah Sandstone, which underlies the volcanics, at least in some areas, is unproven. Although it outcrops at the base of the horst, the Tawilah Sandstone has not been reached by any borehole in the graben.

## Outline Methodology

The analysis of the alluvial Al Hayma aquifer is based on the principle that all water inputs to the valley must equal the outputs, plus or minus any change in groundwater storage:

Rainfall + Runoff =     Outflow + Evaporation & Evapotranspiration
+ Abstraction for City +/- Change in Storage

Runoff is that into the valley from the surrounding hills and, like outflow, may have surface and sub-surface components. Evaporation and evapotranspiration mainly comprises that from soil and from plants.

Consumption by the population comprises less than 0.2% of the total water budget. Change in storage is the net result of subsurface inflow and outflow and infiltration both from rainfall events and "re-infiltration" from irrigation. The change in storage is observed through hydrographs.

In order to address the five points listed above, a number of models have been developed to simulate the terms in the water-balance equation so as to reproduce the observed decline in water levels, that is, change in storage in the Al Hayma valley, over the period abstraction for the city has been taking place. This constitutes the period 1983-1995. Because each of the terms have varying margins of error, the rationale behind the modelling has been to eliminate scenarios and assumptions which cannot reproduce the historical trend.

Although it is not possible to arrive at a unique solution, the error margins of individual variables when considered together (sensitivity analysis) still permit an evaluation which addresses the five points. The methodologies for determining the inputs to the models are described in each of the following sections;

1.   the origin of the rainfall data,
2.   runoff to the Al Hayma valley,
3.   evapotranspiration from the valley with changing irrigation and cropping practices, and
4.   attempted reproduction of the observed hydrographs by a series of groundwater models and prediction of the aquifer recovery period.

Table 2.1 gives a summary description of the function, inputs and outputs of each of the models.

## Rainfall

The area lying within the previous political boundaries of North Yemen may be roughly divided topographically into three zones running north-south: the western coastal strip (or Tihamah) flanking the Red Sea, the central highlands, and the eastern slopes descending towards the desert of the Rub' Al Khali. This topographic division largely determines the rainfall distribution, with very little rainfall in the Tihamah and the eastern desert. The central highlands receive most of the rain, which increases southwards, reaching a maximum of over 1000mm/yr in the vicinity of Ibb, only 35 to 40 km north of Ta'iz.

The northward movement of the inter-tropical convergence zone, sometimes referred to as the Red Sea convergence zone (Gun and Abdulaziz, 1995; 21-22 and WAPCOS, 1996; 2.1), in early summer (April-May) and its southward movement in late summer (August-September) results in increased precipitation at these times (Williams, 1979; 3-6). Precipitation during this bimodal rainy season is in marked contrast to the dry season (October-March) of prevalent north-east trade winds, and in many areas, including Ta'iz, to the noticeably drier months of June-July (Figure 2.3).

**Table 2.1 Summary description of the function, inputs and outputs of each of the models**

| Function | Inputs | Outputs |
|---|---|---|

*SCS / TS-HWC Runoff Model*
*Daily data from 1983 to 1995.*

| Function | Inputs | Outputs |
|---|---|---|
| To generate runoff volumes to Al Hayma. | Airport Daily Rainfall, SCS Land Use Types. | Daily flows from tributaries and flanks to be applied to central wadi course and edge areas of Al Hayma (as mm). |

*Penman-Monteith Evapotranspiration Model*
*Daily data (1983 to 1995)*

| Function | Inputs | Outputs |
|---|---|---|
| To estimate the amount of water required by each crop from abstraction above and beyond that provided by rainfall and runoff and, where water input from the latter two exceeds demand, to estimate the infiltration. | Airport daily mean dry and wet bulb temperatures, sunshine hours, Airport and Usayfra wind run. z-d and crop heights. Output from Runoff. | Net infiltration and abstraction for locations receiving central spates, flank spates and just rainfall for the appropriate cropping pattern ($m^3$ per 100m x 100m grid square per season. Dry season Oct-Feb, wet Mar-Sept). |

*Steady State Groundwater Flow Model,*
*1976 pre-development calibration*

| Function | Inputs | Outputs |
|---|---|---|
| | Aquifer geometry, hydraulic conductivity, subsurface tributary and outlet constant heads, net recharge derived from evapotranspiration model. | Calibration of flows and heads. |

| Function | Inputs | Outputs |
| --- | --- | --- |

### Transient Water Balance Model a)

| Function | Inputs | Outputs |
| --- | --- | --- |
| To run a mean rainfall year (1987) five times, incrementing drawdowns to examine whether the scenario being considered could approach matching the observed hydrographs. | Satellite image determined irrigation areas for 1986. Output from evapotranspiration model for 1987. | Calibration of hydrographs. Assessment of flows. |

### Transient Water Balance Model b)

| Function | Inputs | Outputs |
| --- | --- | --- |
| To match the complete observed hydrograph record from 1983 to 1995. | Satellite image determined irrigation areas for 1986 and 1995. Output from evapotranspiration model for 1983 to 1995. | Calibration of hydrographs. Assessment of flows. |

### Recovery Model

| Function | Inputs | Outputs |
| --- | --- | --- |
| To predict how long it would take the aquifer to recover to approximate pre-development levels if the city continued to abstract as in 1996 and there was no irrigation. | Satellite image determined irrigation areas for 1995. Output from evapotranspiration model for 1987. | Predictive hydrograph. |

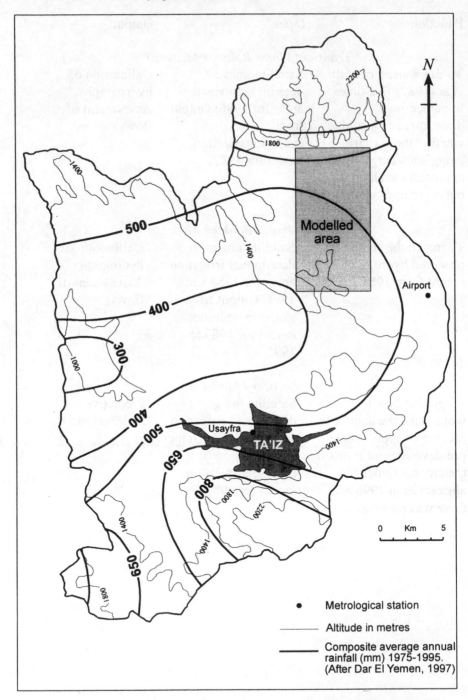

**Figure 2.2  Rainfall distribution – Upper Wadi Rasyan catchment**

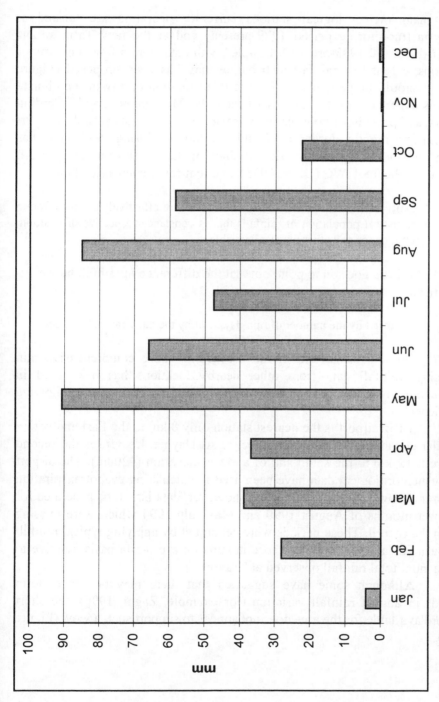

**Figure 2.3 Average monthly rainfall (Ta'iz Airport, 1983-1995)**

The main meteorological stations currently operating are located at Usayfra (monitoring period 1979-present) and at the new Ta'iz airport (1976-1979 and 1983-present). Coupled with data from a few, more minor stations, a tentative rainfall distribution map has been proposed (Figure 2.2). It should be pointed out that the interpretation of any of the climate data is dependent on the quality of the data. Measurement and recording errors, and periods without readings are present in the data sets. Also, apart from a few months during 1977 and two months during 1989, there has been no monitoring of climatic variables in the Al Hayma valley itself. However, the TS-HWC (DHV, 1993) note that for Yemen as a whole:

> Daily rainfalls observed at any one station are effectively samples from a statistical population of rainfalls that is constant irrespective of position, altitude, or any other physical variable.

and DHV 1993 goes on to point out that the difference in rainfall between locations:

> is caused by the number of rain-days, not by the daily rainfall amounts.

On this basis, and in the absence of more comprehensive rain gauging, rainfall data from other nearby locations has been used in estimating the runoff and evapotranspiration described in the following sections.

Ta'iz airport is the nearest station only 8 km to the East and with a similar elevation (1450m) and aspect to Al Hayma. Usayfra is the second nearest, 12 km to the south and of a lower elevation (1200m). The airport daily meteorological data have been used in calculating evapotranspiration and runoff over the modelled period, however, data had to be generated for the wet months of August 1990 and May-July 1994 which were missing from the record. These periods were generated by applying typical rainfall frequencies and amounts for those months on a pro-rata basis, relative to the annual total rainfall observed at Usayfra.

Although some have suggested that there may be longer term trends in annual rainfall variation (for example, Zagni, 1996), the short record available for the area does not appear to support such a view (Figure 2.4).

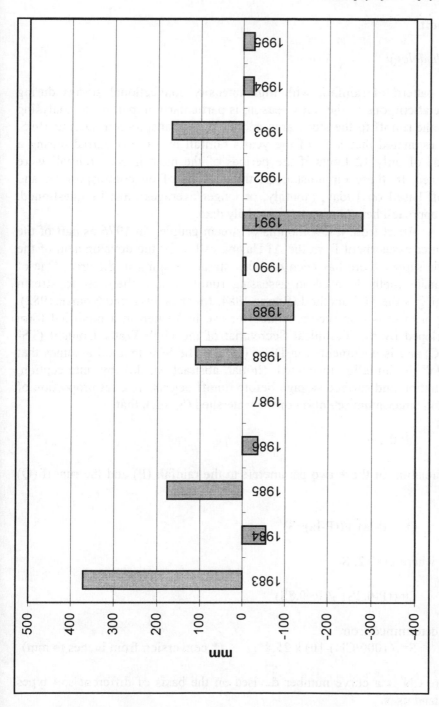

**Figure 2.4 Annual rainfall - variations from the mean (Ta'iz Airport)**

## Runoff

*Methodology*

The pattern of rainfall, with high intensity convectional storms during some afternoons in the rainy season, is particularly important in analysing both the runoff to the study area and the evapotranspiration. During 1995, it is estimated that 85% of the year's rainfall in Ta'iz occurred during a period of only 12 hours if the periods of the most intense rainfall were summed. In these circumstances, the analysis of evapotranspiration and runoff based on 10day, monthly, or longer averages must be questioned. The approach here has been to use daily data.

Apart from a few months of stream gauging in 1976 as part of the resource assessment from the Al Hayma valley for the development of the public supply there has been no other stream gauging in the area. One of the main methods used in assessing runoff where there is no stream gauging is the SCS method (Rango,1984, Mockus,1972 and Noman,1982). This method is also recommended for use in Yemen in a modified form developed by the Technical Secretariat of the High Water Council (TS-HWC) and is documented in DHV (1993). The SCS method assumes that rainfall is initially abstracted (Initial abstraction, Ia) by interception, infiltration and surface storage before runoff begins, as a set proportion of the total maximum retention of the watershed (S) such that:

$$Ia=0.2 \ S$$

The relation of these two parameters to the rainfall (P) and the runoff (Q) is:

$$Q=((P-Ia)^2)/((P-Ia)+S)$$

and, where Ia=0.2, S:

$$Q=((P-0.2S)^2)/(P+0.8S)$$

S is determined from
$$S=((1000/CN)-10) \times 25.4*, \qquad (* \text{ conversion from inches to mm})$$

where CN is a curve number devised on the basis of different soil types and land uses.

Although originally developed on Eastern USA catchments, TS-HWC (DHV, 1993) have suggested values of Ia and CN applicable for Yemen. In order to test the relevance of either method, monitoring of wadi flows during storms in Ta'iz area on the slopes to the south of the graben has been carried out. Although the monitoring was not possible in the wadis discharging to the Al Hayma valley (located on the slopes to the north of the graben), they are similar catchments in terms of topography, altitude, land use, soils, rainfall amount, intensity and frequency. The difference in aspect does not seem too important since the local prevailing winds during the rainy season are westerlies, and sometimes easterlies.

During the rainy seasons of 1995 and 1996, thirteen wadi flows from ten rainfall events were monitored. Flow estimates comprised rudimentary measurement of the stream cross-section and the velocity of floating objects at regular intervals during the storm. Locations were chosen where the wadi bed profile was relatively smooth and constant. Allowance for drag was made by taking a mean velocity of 75% of the observed surface velocity (van der Gun pers comm.). Rainfall measurement was by rain gauge mostly located within the catchment. Due to the absence of other rain gauges, it was not possible to carry out rainfall averaging methods over the catchments, such as Thiessen polygons, although it should be noted that the catchments are adjacent and of similar aspect and elevation. A representative hydrograph is shown in Figure 2.8.

The curve numbers were converted for antecedent moisture condition according to the previous 5 days rainfall and the season, according to Tables 2.4 and 2.5 in DHV, 1993.

The observed and calculated wadi runoff / rainfall ratios are contrasted in Figure 2.5. The rainfall events were typically intense, with a distinct start and finish and the lag time between commencement of rainfall and the peak flow was of the order of 15 to 20 minutes. Although the ratio determined by calculation was greatly different from the observed data in some instances, these instances tended to be for the lower rainfall events where a greater margin of error might be expected due to the rain gauge being distant from the storm centre. For the purpose of adopting a method to simulate runoff, it is more important that calculated and observed runoff are closer for the higher rainfall events, as was the case. On the whole the TS-HWC values were closer to the observed than the SCS ones. This is thought to be because land type descriptions, and therefore land type areas, in the TS-HWC method more closely approximate the field condition. The TS-HWC CN and Ia values have been used in preference to the SCS ones in calculating runoff to Wadi Al Hayma.

[The SCS model and its application by TS-HWC are regarding runoff. The use of it here is to determine runoff from the surrounding wadis and slopes to the Al Hayma valley. Perhaps, more correctly, this should be termed runon, however here it is referred to as runoff. Subsurface outflow through the Miqbabah gorge is referred to as "outflow" in this chapter.]

**Table 2.2  Curve numbers derived from the SCS method and those proposed by TS-HWC as applicable to Yemen**

| [a]Land Use Classification | [b]SCS Description | [c]SCS Soil | [b]SCS Curve Number | [b]TS-HWC Description / Type / Ia value | TSHWC Curve Number |
|---|---|---|---|---|---|
| Uncultivated steep slopes | Bare Ground | D | 90 | Steep slopes with bare rocks / P1 / 0.15 | 90 |
| Uncultivated lower slopes | Bare Ground | C | 88 | Low slopes with bare rocks or thin soils /P2/0.2 | 85 |
| Terraced steep slopes | Row Crops Contoured and Terraced | B | 72 | Terraces on slopes / P5 / 0.3 | 65 |
| Cultivated Wadi Floor | Row Crops Contoured and Terraced | A | 64 | Terraces in wadi beds or on plains / A2 / 0.3 | 65 |
| Urban Area | Residential | B | 74 | Flat areas with impermeable soils / P4 / 0.2 | 75 |

[a] adapted from Dar al Yemen, 1997
[b] DHV, 1993 and Rango 1984
[c] matched with soil types given in Dar al Yemen, 1997

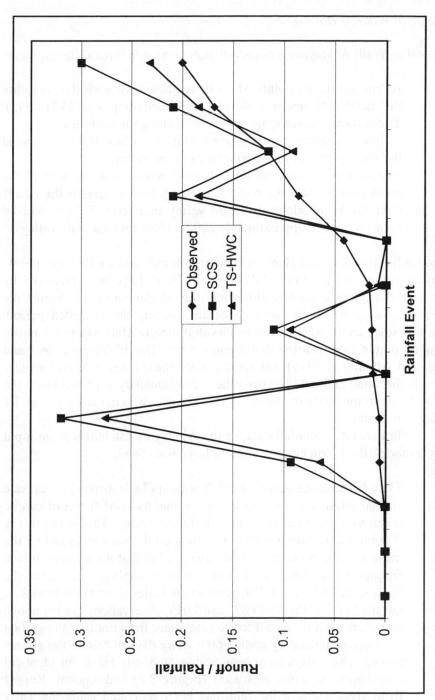

**Figure 2.5  Comparison of observed and calculated runoff / rainfall coefficients**

*Runoff to Wadi Al Hayma*

Runoff to Wadi Al Hayma is observed as occurring in three different ways:

1. Inflow at specific points where there are known wadi flows (within the period of resource development/modelling viz. 1983-1995). These form a linear spate flow source along the wadi bed.
2. Inflow at specific points where major subsurface inflows occur at the entry points of major tributaries to the valley.
3. Runoff from a large number of small wadis on either side of the main valley where the water harvesting system spreads the runoff over the fields along the main valley perimeter. This is said by locals to extend approximately 200 to 300m into the main valley.

The wadis, Rahabah and Hama'ir discharge minor surface flows to the Al Hayma valley. Wadis Tanif, Ja'ashin, and Hajib have been observed by locals to discharge as surface flows into the Al Hayma valley, though the flow frequency has been noted to decline during the modelled period. These wadis receive water from the elevated high rainfall area to the north. It is estimated from rainfall distribution maps (Dar El Yemen, 1997 and Gun and Abdulaziz, 1995) that rainfall over these areas is approximately 60% higher than in the Al Hayma valley. The boundary is placed along the E-W fault at the northern limit of the graben, which marks the major change in slope.

In generating rainfall data for the Al Hayma catchment as an input to the model, the following assumptions have been used:

1. That, because the nearest rainfall station (Ta'iz airport) occupies a similar altitude and aspect to Al Hayma, its rainfall record can be taken as representative of the Al Hayma valley. This is not to say that rainfall occurred in the catchment on the same days and by the same amount as occurred at the airport, but that the monthly totals, frequency and duration of storms will be similar. This forms the basis of the daily rainfall applied to the valley floor in the model.
2. On the basis of the TS-HWC and DHV observations quoted above regarding rainfall, runoff to the catchment from north of the graben has been generated by adding 60% more days of rain to the airport record. These days have been added randomly but in an identical distribution in terms of season (Figure 2.3) and amount. Runoff from areas south of the fault has been generated using the Ta'iz airport rainfall record.

In addition, runoff observations in the catchments around Ta'iz suggest:

1.       that daily rainfall below 6mm (Figure 2.6) should not contribute to runoff (DHV 1993;5) and therefore should be excluded, and
2.       for catchments of the order of tens of km$^2$, that a maximum runoff/rainfall ratio of 10% occurs (Figure 2.7).

In order to assess the validity of the last point, as part of the model sensitivity analysis, two variants have been considered:

1.       runoff without a maximum cutoff, and
2.       a runoff maximum of 10% of the rainfall.

The main reason for the decline in the frequency and amount of spates reaching Al Hayma in recent years is considered to be the significant increase in groundwater abstraction for agriculture in Wadi Hajib, and the associated change in land use, and for the Ta'iz municipal supply below Wadi Ja'ashin (viz. Wadi Al Minqa'ah). Although this is part of the point the model is attempting to analyse, in the case of Wadi Al Hayma, the observation by locals of declining spates into Wadi Al Hayma must be built into the model inputs.

      The model developed by SCS also assumes that the whole of the catchment received the rainfall of that day and that the antecedent moisture condition was the same for the whole catchment. In reality this is not the case and some of the runoff absorbing zones DHV (1993; 16) will do exactly that, and absorb the runoff from other parts of the catchment before it reaches the exit point. The curve numbers "lump" together an overall average runoff/absorbtion property for the catchment being considered. In summary, effectively the development of the water resources has increased the water absorbing capacity of the catchments with time and changed their curve numbers. Because no calibration for this is available, the calculated runoff from wadis Tanif, Ja'ashin and Hajib to Wadi Al Hayma in this analysis have had to be altered so that they occur only during maximum spate periods. The largest spates in the field occur in August and May. Runoff in the model to Wadi Al Hayma from these wadis has only been permitted in August, when the model usually generates one or two spate periods of a few days.

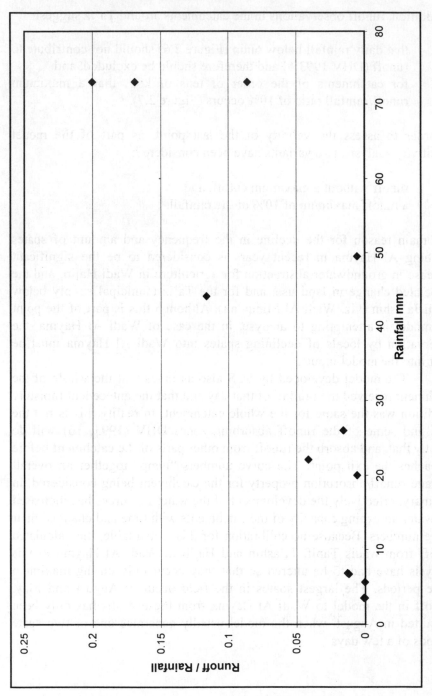

**Figure 2.6  Runoff threshold from individual storms in Ta'iz**

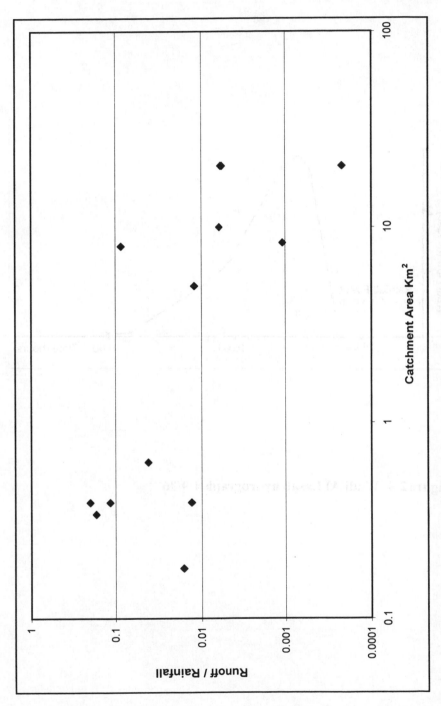

**Figure 2.7 Runoff / rainfall vs. catchment area relationship from individual rainfall events**

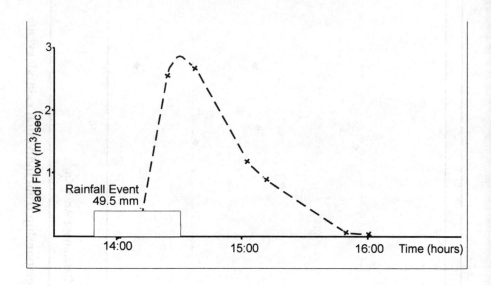

**Figure 2.8  Wadi Al Lasab hydrograph 4/9/96**

**Table 2.3  1983-1995 mean runoff (m³/day) to the Al Hayma valley from the main tributaries and the flanks as determined by the different methods used in the runoff model**

| Runoff Type: | 1 & 2 | 1 & 2 | 1 & 2 | 3 | 3 | |
|---|---|---|---|---|---|---|
| Wadi: | Tanif | Ja'ashin | Hajib | West Flank | East Flank | Total* |
| Model Variants | | | | | | |
| a) SCS-TS-HWC | 5000 | 5600 | 4100 | 2300 | 1400 | 21200 |
| b) 10% max Q/P | 1700 | 2000 | 1500 | 800 | 500 | 7500 |

*also includes some minor tributaries

These values are compared with those determined from the sensitivity analysis of the steady state groundwater flow model below.

**Evapotranspiration**

*Choice of Evapotranspiration Model*

Doorenbos and Pruitt (1977; Table 1) and more recently, Shuttleworth (1990, cited in Wallace, 1991; 141) have summarised some of the methods used to determine evapotranspiration rate. The methods may be divided into direct methods (using lysimeters, evaporation pans), and a range of highly empirical to very complex physically-based formulae. Each method has its advantages and drawbacks, but underlying all of them is the problem of Actual Evaporation falling below Potential Evaporation estimated by the method due to restricted uptake of water by the plant root system (an area still not clearly understood, Monteith, 1991; 20). Essentially the methods of measuring evapotranspiration directly require equipment not generally available (or in some states, such as Yemen, poorly maintained), the empirical methods have very limited applicability, and the physically-based methods require equipment which, although used in a research context, is not found in routine meteorological station monitoring. The best compromise of applicability, accuracy and data requirement is in the less complex physically based formulae.

Since the FAO recommendation of 1990, the Modified Penman method of Doorenbos and Pruitt (1977) has been superseded by the Penman Monteith method (Monteith, 1965) as a basis for calculating reference crop evapotranspiration. Both these methods involve an energy input term and an aerodynamic term. The latter method uses a more physically-based estimate of resistance to energy and water vapour flux, simulating evapotranspiration as though it were from one "big leaf", representing the crop canopy.

The changes in crop water requirements, and hence "losses" to evapotranspiration and evaporation from plant and soil as the plant develops were "built into" the Modified Penman method via empirical crop factors rather than through the physically-based resistances. The crop factors also relate the crop water requirements for various crops to that of a "reference crop" of 8 to 15cm grass not experiencing water shortage. More recently, the "big leaf" concept has been questioned and the importance of evaporation from bare soil in between the plants which significantly changes the aerodynamic properties of the "field" has been recognised. This has led to a proliferation of resistance terms in the formulae to cover soil, stomatal, canopy and aerodynamic resistances (Shuttleworth and Wallace, 1985; 842).

Complex multi-layered models applicable to detailed research at the individual field scale, and smaller, have been developed, but these methods require sophisticated research equipment for the lab/field measurements (Wallace, 1991; 140).

## FAO Penman Monteith Model Parameterisation

The derivation of the Penman Monteith equation for the estimation of evapotranspiration is described in Monteith (1965, 1991), and in Monteith and Unsworth, (1990; 247 et seq.). The FAO Irrigation and Drainage Paper No. 56 "Crop evapotranspiration" incorporates crop coefficients which aggregate the physical and physiological differences between crops relative to a reference crop of standard surface resistance (Allen et al., 1998; iii).

Daily measurements of mean dry bulb temperature, mean wet bulb temperature, wind speed and sunshine hours were provided from the Ta'iz airport meteorological station. Saturation Vapour Pressures were calculated from the wet and dry bulb temperatures using the method described in Abbot and Tabony (1985).

Values of location-specific variables are given in Table 2.4. Other parameters and their units used in the Penman Monteith equation are described in various places such as Monteith and Unsworth (1990; 247) and Shuttleworth and Wallace (1985; 842).

**Table 2.4 Values used for the location-specific variables/constants of the Penman Monteith equation and the basis for their selection in the Ta'iz context**

| Parameter | Description | Value / Unit | Basis |
|---|---|---|---|
| alpha | albedo | 0.23 | Allen et al 1998; 23 |
| a | Constants in equation for atmospheric emissivity | 0.34 | Rhebergen & van Waveren 1990 for Yemen highlands cited in Al Derwish, 1995 |
| b | ditto | 0.044 | ditto |
| a' | Constant in equation for net long wave radiation | 0.2 | Al Derwish, 1995 |
| C | Surface soil heat flux = C (net radiation) | 0.3 | Fuchs and Hada, 1972 cited in Al Derwish, 1995 |
| P | Atmospheric Pressure +/- 5 mbar | 858 mbar | Ta'iz airport measurements |
| $r_s$ | Surface resistance to heat and water vapour fluxes for well watered vegetation | 70 sec/m | Allen et al 1998; 22 |

However, the evaluation of $r_a$, the aerodynamic resistance to heat and water vapour flow also requires site-specific inputs:

$$r_a = [\ln((z-d)/z_o)/k^2 u]^2$$

Where, from Monteith and Unsworth, 1990:

k       is the von Karman constant = 0.41

u       is the wind speed (m sec$^{-1}$)at the measurement point (typically 2m above ground level)

z       is the reference height, that is, the height of the measurement point above the ground

d       is the equivalent height for the conservation of momentum, such that:

$d+z_0$     is the height of zero wind speed

Typical relationships (ibid; 117 and Allen et al, 1998; 21) are:

$d=0.65$ x crop height
$z_0=0.123$ x crop height

Independent checking of wind speed in Ta'iz indicated 50% lower wind speeds than those measured at the airport and were closer to those measured at the Usayfra meteorological station. Ta'iz airport is more exposed than the Al Hayma valley or the Usayfra site. Although wind measurements at both are at 2m height, the Usayfra site conforms better to the intended application of the Penman Monteith method. Modelling has compared both airport and 50% airport wind speeds.

*Soil Moisture Content: Model Methodology*

The absence of data regarding soil moisture content with depth at different times since rainfall or irrigation precludes the use of a zero flux plane model. The simpler FAO readily available water (RAW) model (Allen et al, 1998; 161 et seq.), more commensurate with the available data, is used here. For the summer crops, it is assumed that for the medium silty loam soils of the Al Hayma valley (Dar El Yemen, 1997; 27) the total available soil moisture content is 14% (Smith, 1992; 36 and Allen et al, 1998; 144). Applying the effective rooting depth (ibid. 61 and Allen et al, 1998; 163) for each crop type at each growth stage gives the RAW for the crop.

*Irrigation*

Current cropping patterns of the major crop types are given in Table 2.5 together with crop coefficients used. Crop coefficients were calculated on the basis of Allen et al (1998; chapter 6).

Rainfall seasonality results in maximum irrigation occurring during the dry winters and supplementary irrigation in the summer. In the case of winter irrigated crops, it has been assumed that the crop water requirement is met, and the FAO Penman Monteith method has been used in calculating the reference crop evapotranspiration.

It should be noted that the "rainfall" used in evapotranspiration calculations is the recorded rainfall, whilst irrigation is an effective irrigation. Effective irrigation is defined as the water pumped for irrigation less that portion which recirculates to the aquifer, which can be, and has been, ignored in terms of the total water balance. As an approximation it is assumed that evaporation from channels is similar to that of the evapotranspiration from the crops they irrigate and is proportional to the area they occupy. The estimation of irrigated areas by satellite imagery discussed above includes the channel area. This is also effectively one of the assumptions in the image analysis; that the growth of natural vegetation around the channels contributes to the image and the registering of an "irrigated pixel". This assumption together with the occurrence of high infiltration rates result in small evaporative loss contributions to field application efficiency (though there are large infiltration "losses") and no further calculation for irrigation losses has been made.

**Table 2.5 Cropping patterns, length of season (days) and crop coefficients ($K_c$ini, mid, end) for major crop types**

| Winter Irrigated | Summer Supplementary Irrigated | Summer Rainfed |
|---|---|---|
| Potatoes: 120-125 days, (0.5,1.2,0.8) | Maize: 140-145 days, (0.7,1.3,0.58) | Sorghum / Millet: 125-130 days, (0.7,1.1,0.5) |
| Tomatoes: 180 days, (0.6,1.2,0.815) | Sorghum / Millet 125-130 days, (0.7,1.1,0.5) | |
| Maize: 145 days, (0.6,1.275,0.57) | Qat*: 365 days, (0.65) | |
| Qat*: 365 days (0.65) | | |

* A cash crop bush, similar to privet, the leaves of which are chewed by many Yemenis each afternoon because of its amphetamine / stimulant properties.

*Summer rainfed crops* The FAO RAW methodology has been used with both irrigated and rainfed crops. Inputs from rain and daily abstractions by crops from the tank are dealt with as described in the previous paragraph except that the only input is rainfall.

*Winter irrigated crops* The rare occurrences of rainfall in the winter have been evaporated from the non-irrigated areas (that is, bare soil areas) in the model at a rate proportional to the reciprocal of the square root of time after an initial two day period of evaporation at the potential rate for wetted soil (Monteith, 1991; 14 et seq.) which is assumed to be at 90% of the potential rate for an open water surface calculated using the Penman method (Wilson, 1990; 51). It is assumed that the RAW for bare soil is 35mm (equivalent to 14% porosity over 0.5m depth, which compares reasonably with Smith, 1992; 61, and Monteith, 1991; 15).

Although these are very rough approximations, the number of uncropped periods in which there are more than two successive rainfall days are very few. Thus large changes in the assumed values of the variables mentioned above have a small effect on the volume of water available for infiltration in the model.

*Supplementary irrigated crops*  With supplementary irrigated crops, it is important to assess the amount of water from the farmer's contribution via irrigation and that from rainfall and runoff. These amounts are summed separately in the evapotranspiration model.

*Qat irrigation*  Qat crop water requirement calculation is problematic. Qat has been considered analogous to citrus fruit in calculations (Zagni, 1996), however, local irrigation practices indicate that the plant can survive on much smaller quantities of water and in the dry winter season almost "hibernates" if it is not irrigated. Farmers will typically irrigate in order to obtain one crop in the lucrative winter months. This usually comprises three irrigation turns at around 10 day intervals with a maximum of 10 to 15 cm application each turn, applied directly to a basin surrounding each individual bush.

It is unclear if qat receives supplementary irrigation in the summer. Many farmers comment that they obtain three crops per year, one in winter and two in summer.  This would coincide with the bimodal summer rainfall. However it is unlikely that with irrigation facilities to hand (which nearly all the Al Hayma valley qat farms have) no supplementary irrigation takes place.

From discussions with farmers, it has been assumed in the model that three crops per year are irrigated / supplementary irrigated; one in winter, one in early summer and one in late summer. It has been assumed that three turns are applied to the winter crop, and that a significant rainfall event in both of the summer rains triggers the farmer to aim for a crop which then needs a further one or two irrigation turns, depending on whether the date of the next rain is before or after the 10 day irrigation interval.

*Results* Evapotranspiration by different crops (Table 2.6) has been calculated using wind measurements from both Ta'iz Airport and 50% of that rate, as discussed above.

$ET_o$ values calculated by the pre-FAO 56 Penman-Monteith method in other studies have ranged from 5.8 (Zagni,1996, Ta'iz Airport), to 4.0 (Dubby and Taher,1998, Al Hayma).

**Table 2.6  Mean crop evapotranspiration (mm/day) calculated for the full season**

|  | Airport Wind, | 50% Airport Wind |
|---|---|---|
| Supplementary irrigated sorghum/ millet | 5.26 | 4.58 |
| Rainfed sorghum/millet | 2.05 | 2.00 |
| Summer maize | 5.89 | 5.15 |
| Winter maize | 3.53 | 3.26 |
| Potatoes | 2.67 | 2.52 |
| Tomatoes | 3.65 | 3.38 |
| Qat | 2.20 | 2.14 |
| $ET_o$ | 4.76 | 4.24 |

*Irrigation Trends*

In order to assess the evapotranspiration throughout the modelled period, it was particularly important to determine the extent of irrigation activity over this period. Interviews with local farmers suggested that irrigation activity in the central portion of the Al Hayma valley peaked in the mid-80's, after which, according to the farmers, increasing abstraction by the water authority for the Ta'iz municipal supply began to deplete the aquifer significantly. Subject to availability, financial budget, and the absence of cloud, Landsat TM data were selected for the peak central valley irrigation period (1986) and for recent times (1995). Summer (June and September) and winter (January) data were obtained for both dates so that rainfed (summer image) and irrigated areas (summer and winter images) could be distinguished. The recent data facilitated field checking of current water use during the summer and winter seasons of 1996. From the Landsat data, various images were produced by SOAS researchers undertaking projects in remote sensing. Three main assumptions are involved:

1.    In January, all vegetation is irrigated.
2.    Pixels with a high response in TM4 (Very Near Infrared) and a low response in TM3 (Red) are vegetation.
3.    Pixels classified as "irrigated" are 100% irrigated – that is, there are no part-irrigated pixels (pixels cover a horizontal land area of 30m x 30m).

Four main images were produced. Although Bands 1, 2, 3, 4, 5 and 7 were available, the two vegetation indices (NDVI and TVI) use TM4 and TM3 bands only, the latter taking into account soil brightness. The other two methods, Principal Component Analysis and Tasseled Cap Analysis are statistical analyses which look at interband correlations. All four techniques produce different estimates of irrigated area, the variation being +/-20% of the mean. The main reason for the difference is the thresholding process. A threshold digital number is selected manually to a level which selects as much of the area where vegetation is thought to be without selecting too many pixels outside this area which are considered noise. Although the statistical methods are considered more "scientific", field checking suggested they produced too dense a level of irrigation. The NDVI appeared closest to the field situation and the TVI image appeared too "sparse". These observations could not be verified by detailed field/image correlation however. The NDVI image was selected for identifying irrigated area for input to the water balance model. The total irrigated area in the modelled part of the Al Hayma valley by the NDVI method was $3.7km^2$ for 1986 and $4.3km^2$ for 1995 indicating a 16% increase in irrigated area over the modelled period as well as a more obvious huge upstream shift in irrigation.

The summer images and the field checking indicate that the whole valley is cultivated in the summer, thus areas not cultivated during the winter are assumed to support rainfed sorghum/millet during the summer. The proportions of these crops applied to the irrigated areas identified by the satellite imagery are used to calculate the areal distribution of evapotranspiration with time that forms an input to the groundwater models.

## Model Checks

*Modelled and observed irrigation frequencies*   The extent to which irrigated crops fall below the threshold (Allen et al, 1998;169) is a useful check on the model, since, assuming farmers recognise when the crops need irrigating, this should happen rarely and to an insignificant amount.

**Table 2.7 Approximate irrigated crop portions in the Al Hayma valley derived from discussions with farmers and used in the evapotranspiration model**

| Area in Al Hayma | Summer (Supplementary) | Winter |
|---|---|---|
| Top 1986 & 1995 | 50% Qat, 25% Sorghum/Millet, 25% Maize | 50% Qat, 50% Maize |
| Upper 1986 & 1995 | 50% Sorghum/Millet, 37.5% Maize, 12.5% Qat | 50% Potatoes, 25% Maize, 12.5%Tomatoes, 12.5% Qat |
| Central / Lower 1986 | 50% Sorghum/Millet, 50% Maize | 75% Maize, 12.5% Potatoes, 12.5% Tomatoes |
| Central / Lower 1995 | 50% Maize, 25% Qat, 25% Sorghum/Millet | 37.5% Maize, 25% Qat, 25% Potatoes, 12.5% Tomatoes |

Table 2.8 suggests that the combination of irrigation intervals noted in the field (and used in the model) and soil moisture availability / rooting depths advised by Allen et al (1998) and Smith (1992), that were used in the model, appear reasonable. Potatoes appear either to be irrigated too infrequently or the suggested soil moisture availability / rooting depths are too low, or both. Certainly the limitations for irrigation imposed by the system of turns does result in periods of stress especially in the summer if rains do not fall in adequate amount, and, more importantly, frequency.

*Total amount of water needed for irrigation* The amount of water which the model predicts should be applied by irrigation provides a rough check for comparison with the amount the farmers could use based on the number of wells and the typical yields. The UNDDSMS well inventory located 123 dug wells and 5 private boreholes in the modelled area, many of which are now dry. The anticipated yield of these particularly during the early stages of the development of the resource would have been of the order of 6-10 lit/sec suggesting a maximum total possible yield of 34,000 to 56,000 m$^3$ per 12-hour day.

**Table 2.8 Proportion of the growing season of each crop in which the soil water content falls below the threshold in a mean rainfall frequency year as determined by the evapotranspiration model**

| Irrigated / Supplementary Irrigated Crops | % of season below threshold | Maximum deficit below threshold (mm) |
|---|---|---|
| Sorghum / Millet | 8 | 10.1 |
| Winter Maize | 10 | 9.2 |
| Summer Maize | 8 | 10.0 |
| Potatoes | 50 | 18.8 |
| Tomatoes | 18 | 10.6 |
| Qat | 42 | 55 |

The irrigation demand suggested by the evapotranspiration model for the total irrigated area in the valley (discussed later) could, in broad terms, have approached a maximum of 15,000 m³/day in the dry season and 6,000 m³/day in the wet season. It would therefore appear that there was sufficient pumping capacity to provide irrigation for the demand indicated by the evapotranspiration model even allowing for small irrigation efficiencies.

Irrigation rates predicted by the evapotranspiration model and also runoff quantities added to the rainfall and applied to the runon areas in the transient water balance model require high infiltration rates if the fields are not to be flooded for long periods. The mean measured infiltration rates (Dar El Yemen, 1997; 28) are 180mm in a three hour period and can approach twice this amount. Typical furrow depths observed in the field are of the order of 150mm. This is adequate to infiltrate nearly all of the model-generated irrigation depths. Even excessive over-irrigation or the maximum, 400mm model-generated irrigation depths, could therefore be infiltrated within three hours. The maximum runoff (spate) depth from the entire thirteen years of data is 1m. The high observed infiltration rates could also cope with this amount within a day especially when it is considered that the wadi bed materials will have even higher rates.

**Groundwater Flow Modelling**

In this section groundwater flow modelling attempts to match the observed hydrographs. A unique solution of the water balance equation is not possible due to error margins in the variables. However the five aims mentioned at the beginning of this chapter can still be addressed within those error limits.

*Steady-State Groundwater Flow Model*

The first step towards matching the observed hydrographs involved attempting to reproduce the groundwater heads when the Al Hayma alluvial aquifer was essentially undeveloped. The widespread development of well construction and pump installation was triggered by the flow of remittances from Saudi Arabia, which began in the mid-1970s (Figure 2.12). A groundwater head map was produced from 1976 data as part of the investigation into the Al Hayma valley as a supply source for the city of Ta'iz (Leggette et al, 1981). The 1976 head distribution was matched by constructing a steady-state model using the GWVistas-Modflow software. The development of the inputs to the model is described below.

Aquifer geometry was determined from the extent of cultivated land on the alluvial floor of the valley indicated by the Directorate of Overseas Surveys 1981 maps and the satellite images of the area. The base of the aquifer was determined from borehole information and from the results of a resistivity survey carried out by (Leggette et al, 1981) and matched to the borehole data. The survey and boreholes were located in the central part of the valley and the lower (Miqbaba) part. The level of the base of the aquifer in the upstream, northern part of the valley was estimated by extrapolation. The valley was modelled as comprising one aquifer (the alluvium) because the volcanics are much less permeable (the mean permeability of the volcanics is around 300 times smaller than that of the alluvium, Dar El-Yemen, 1997; 91).

Permeability of the alluvium was determined by those who conducted borehole pumping tests at the time of their construction  by Jacob and Theis recovery analyses. The occurrence of some discontinuous silty and clayey bands presumably led them to use confined methods, although a semi-confined method may have been more appropriate. No observation well data were available, however, storage data are not required for the steady-state analysis. Values of hydraulic conductivity from the more reliable pumping tests ranged from 333 to 3m/day with a mean of 42m/day and a median of 29m/day.

Wadi inflows were represented in the model by constant heads. Although the software has the facility to model stream flows, the extremely ephemeral nature of the wadi flows precluded their representation by this method in a steady-state analysis, and instead, flows from constant head sources located at the entry points of tributaries to the Al Hayma valley were used in calibration. Constant heads were located between ground level and the base of the aquifer. In attempting to calibrate heads and flows in the model it was found that the heads could not be varied by more than +/- 2.5m and in the case of the major wadis by not more than +/- 1m without the model failing to converge. This relatively tight control meant that the resulting wadi flows did not vary greatly. The occurrence of a marshy area at the narrow outlet of the central part of the valley was modelled as a drain with a water level equivalent to ground level.

Recharge was applied using the rainfall of 1987 (0.00126 m/day, the closest year on record to the mean of the modelled period) to all the cells in the model. The valley perimeter cells also received the extra equivalent rainfall contributed as runoff from the east and west flanks in accordance with farmer's observations. The rainfed crop and bare soil evapotranspiration and evaporation were deducted from the rainfall to give net recharge. As a rough estimate at this initial stage of analysis, the evapotranspiration from a 300m-wide strip potatoes and maize irrigated for three seasons per year and extending the full length of the wadi was deducted in the calculation of net recharge. Again this was an approximation of the farmer's observations regarding the extent of irrigated agriculture prior to significant well construction.

**Table 2.9 Variation in evapotranspiration and runoff inputs to the groundwater modelling**

| Scenario | Model Used to Generate the Scenario |
|---|---|
| Wind velocity data are used from: i) Ta'iz airport, and ii) 50% of this velocity | Evapotranspiration |
| Runoff is calculated by: i) the SCS-TS-HWC method ii) As above, with a maximum of 10% runoff/rainfall ratio | Runoff |

On the basis of Table 2.9, two main scenarios were explored:

1.  a driest case in which evapotranspiration was based on the wind velocities recorded at Ta'iz airport and a runon from the eastern

and western flanks based on the SCS-TS-HWC model but with a 10% maximum runoff/rainfall ratio, and

2.      a wettest case in which evapotranspiration was based on 50% of the wind velocities recorded at Ta'iz airport and a runon from the eastern and western flanks based on the SCS-TS-HWC model.

**Table 2.10  Net recharge (m/day) contributed to the valley as runoff from the eastern and western flanks**

| Case | Flanks (882 cells) | Central Zone (742 cells) | Balance |
|---|---|---|---|
| Dry: Airport wind velocity and SCS-TS-HWC | +0.00018 | -0.00054 | -0.00015 |
| Wet: 50% Airport wind velocity and SCS-TS-HWC with 10% max Q/P ratio | +0.00049 | -0.00041 | +0.000079 |

(+ denotes net water gain, - denotes net water loss):

Groundwater heads and wadi flows were used as model targets. The target head distribution was within 5m of the observed 1976 levels over the full 240m range in head values from the northern to the southern end of the valley. Wadi flow targets were within the range given in Table 2.3. Recharge was varied according to Table 2.10 and intermediate cases were also considered. Wetter cases failed to reach a solution and dryer cases caused the model to "run dry". Permeability was varied to determine the limits in which a solution could be obtained with a water balance of better than 0.01%. The head convergence criterion was 1mm.

**Table 2.11  Steady state model sensitivity**

| Scenario | Dry | Intermediate | Wet |
|---|---|---|---|
| Solution & Balance | OK | OK | OK |
| Permeability m/day | 10-100 | 20-25 | 34-38 |
| Comment on heads and flows | Best when K=15 to 20 | Best when recharge =0.00007 m/day | Flows too high |

A net recharge of minus 0.00007 m/day +/- 15% and a permeability range of 20-25 m/day gave the best fit. Over this range the modelled outflow from Miqbaba varied from 4100 m$^3$/day to 11300 m$^3$/day. For comparison, the average flow from Miqbaba monitored by Montgomery (1975) in 1974 was 5,600 m$^3$/day, and by Leggette et al (1981) was 10,400 m$^3$/day. [Their records had to be extrapolated assuming a declining flow during the dry season.] Leggette et al acknowledge that their readings were in a period when groundwater levels were particularly high (1977-1980), whereas this model was being calibrated against 1976 levels.

**Table 2.12 Range of wadi flows (m$^3$/day) determined by the steady state model for the "best fit" range of recharge and permeability and the "target" flows generated by the runoff modelling (Table 2.3)**

| Wadi Flows | Tanif | Ja'ashin | Hajib | Miqbaba | Total Inflows[b] |
|---|---|---|---|---|---|
| Target From: | 1700 | 2000 | 1500 | 5600 | 6200 |
| To: | 5000 | 5600 | 4100 | [a]10400 | 17500 |
| | | | | | |
| Model From: | 2018 | 1900 | 900 | 4100 | 6034 |
| To: | 4200 | 4000 | 1900 | 11300 | 12300 |

[a] observed flows, [b] including minor wadis

The steady-state analysis suggests that the driest case analysed was "too dry" and the wettest "too wet". The best fit solution lies between. The two combinations of Ta'iz airport wind data with the SCS-TS-HWC runoff method, and 50% reduced wind speed with a 10% maximum runoff/rainfall ratio in the SCS-TS-HWC method also lie between the wettest and driest cases and are considered in the transient water balance modelling.

*Transient Water Balance Model*

Recreating the field observed hydrographs by modelling requires a time-series analysis such as that provided by the transient methods available in the GW Vistas-Modflow package. It was decided, however, not to use this method for two reasons. Firstly, because of the problems of cells "running dry" (as has happened in reality), large head differences occur in adjacent cells during successive iterations, resulting in the model failing to converge. Secondly, abstractors export water to different parts of the basin. Thus, although wells in the deeper, central part of the valley remain operative for longer, the farmers away from the centre import water because their wells have run dry as the aquifer was depleted. In trying to model this, a dry cell ceases to function and cannot receive recharge or lose abstraction even though these processes are taking place in reality. Because of the problems of cells "drying up" and farmers importing water, a model was adopted which summed water inflows and outflows on a basin scale.

Rather than the constant head representation of wadi inflow runoff used in the steady state model, the transient model applied the runoff water to the areas described as receiving spate flows by the farmers. These areas are different for the dry and wet season both in the field and in the model. For the modelled period (1983-1995), the area receiving spates has decreased from that in the 1970s when there had been little groundwater development. In particular, groundwater sources had been developed immediately prior to, and during, the modelled period in the Ja'ashin catchment for irrigation and the city supply, and in the Hajib catchment for irrigation. In Wadi Hajib this has resulted in a notable decline in the number of spates reaching the Al Hayma valley. In Wadi Ja'ashin there was an almost immediate interruption of the perennial flow and only very major rainfall events result in spates reaching the Al Hayma valley today. As mentioned above, in an attempt to model this decline in the number of spates, only the runoff from the wettest month (August) has been used in the model for these two wadis.

Water levels have been monitored for the modelled period in wells clustered in the central Al Hayma basin and the Miqbaba basin. Because the limited areas of water level monitoring coincide with the two areas which have a basin geometry with a lip at the exit, a model comprising the two basins with a small interconnecting channel and a large input channel was developed. It was assumed that the basins were horizontal with a horizontal piezometric surface equivalent to that at the geometrical mid-point of the basin. It was also assumed that the level of the "horizontalised" basin outlet lip is at the same level below the mid-point basin start water level as the actual outlet lip is below the starting water level at the outlet. A third assumption was that any excess water in channels (that is, which has not been used by the crops) drains to the basins. This last assumption is not valid because the water takes time to drain into the next basin and during this period is effectively in storage and available for pumping to the overlying crops during a period of deficit. However, because the time steps are quite long (annual wet season and dry season) any excess will tend to be drained from the channels. In any case, the amount of water which crops in the channel areas needed in the deficit periods has been calculated separately and is discussed below, where it is referred to as the "channel deficit". Changes in groundwater level are caused by volumetric inflows and ouflows, thus outflows from the basins and water levels in the basins are interdependent. By taking into account the changing surface area of the piezometric surface at each time step from the aquifer geometry, water levels and outflows could be calculated iteratively for each time step.

*Transient model A: 1987 x 5*  Before attempting to match hydrographs from the full time series, the rainfall, runoff, agricultural and city abstraction data for 1987 (the year closest to the average rainfall) was run for five years consecutively. Model drawdowns were calculated using the "static" level from the steady-state model. The target five-year drawdown was the average drawdown measured in the observation wells in the larger Al Hayma basin during the first five years of the operation of the city supply. There were three main reasons for choosing the five-year period. Firstly that the period would be long enough for the impact of development of the aquifer to be significant. Secondly it was desirable that the modelling should run from a point prior to abstraction for the city. Thirdly, monitoring in the intervening period was very scant and the five-year period offered the first reliable observed heads to calibrate against.

Although pumping tests may show a semi-confined or even confined response, the draining of the aquifer on a large scale comprises the actual replacement of water by air in the pore spaces and a specific yield value is more appropriate to represent storage. A rudimentary variation in specific yield covering the range indicated by the major soil type (14% to 10%) was used.

**Table 2.13  A summary of the analyses and results of the transient water balance modelling of the Al Hayma basin (1987 x 5)**

| Runoff | Wind Run Data | Specific Yield | Al Hayma Drawdowns (m) |
|---|---|---|---|
| SCS with 10% max Q/P | 50% Airport | 0.1 | 4.2 |
| SCS with 10% max Q/P | 50% Airport | 0.14 | 3.5 |
| SCS-TS-HWC | Airport | 0.1 | 3.1 |
| SCS-TS-HWC | Airport | 0.14 | 2.3 |
| | Target five year drawdown: | | 4.0 |

*Transient model B: 1983-1995* On the basis of the results of model A (1987 x 5) three scenarios were selected for attempting to match the complete 13-year record (Table 2.14). The abstraction record was not complete for this period, however.

**Table 2.14 Three scenarios tested by the transient water balance model (1983-1995)**

| Model | 1a | 1b | 2 |
|---|---|---|---|
| Runoff | 10% max Q/P | 10% max Q/P | SCS-TS-HWC |
| Wind Run Data | 50% Airport | Airport | Airport |
| Specific Yield | 0.1 | 0.14 | 0.1 |

The first two years had to be generated by extrapolating back the yields of wells to their commissioning dates and the last five years of data by extrapolating forwards linearly on the basis of declining yields to the date of their eventual failure or their 1995 yields. Irrigated areas were determined for 1983-1987 from the 1986 satellite image and from 1991-1995 from the 1995 image. The intervening period was taken as a 50/50 mixture of the 1986 and 1995 irrigated areas.

The hydrographs for each method are compared with the target hydrographs for Al Hayma and Miqbaba in Figure 2.9. A mass balance error of approximately +/- 10% might be expected.

Model 2 gave the closest match to the observed data. The other two models "dried up" in 1991. It should be noted that matching the Miqbaba water levels was particularly difficult. Because of its relatively small size and low storage, slight changes in the amount of water flowing from the input channel to the Miqbaba basin produce a large change in head.

The discussion up to this point has focused on understanding the natural water movement processes. In summary, the SCS method modified by the Technical Secretariat of the High Water Council for Yemen appears applicable in this instance. However, it is not possible to determine conclusively whether a 10% runoff/rainfall ratio cut off should be used or not due to the error margins of the variables.

**Who Took the Water?**

Leggette et al (1977; 8-18) estimated a 'total quantity available for potential export' for the city of 9 Mm$^3$/yr +/- 20% and proposed an abstraction of 10 Mm$^3$/yr (and an effective end to irrigated farming) as a 'reasonable objective'. In their estimation, the As Sahlah basin accounted for 2Mm$^3$/yr of this supply. This leaves 8 Mm$^3$/yr for abstraction from the area included in this study (Habir to Miqbaba excluding As Sahlah).

From the steady state analysis it is possible to estimate the impact development of the Al Hayma basin has had since the relatively undeveloped situation of 1976.

If no outflow from Miqbaba were to occur, thus negatively impacting agriculture downstream of this point, the potential yield of the is obtained from the sum of the Miqbaba outflow and the evaporative loss from the marsh, viz.: 3.0 Mm$^3$/yr +/- 15%.

Assuming, like Leggette et al, the complete demise of irrigated agriculture, the potential yield is obtained from the sum of the evaporative loss from the marsh, the Miqbaba outflow and evapotranspiration from crops receiving irrigation in 1976. This yield would be 3.8 Mm$^3$/yr +/- 15%.

3.8 Mm$^3$/yr is only 50% of the 'reasonable objective' of Leggette et al (1977). With hindsight it can reasonably be stated therefore that the failure of the Al Hayma supply scheme was guaranteed at its planning stage.

To assess the relative significance of different human water use activities and to determine the physical causes of the 1995 crisis (and earlier crises), abstractions from model 2 were examined (Figure 2.10). Although the design abstractions for the city supply were never achieved, the over-estimates of Leggette et al. resulted in groundwater mining from the outset. At the commencement of abstraction for the city in 1982-1983 irrigation activity alone in Al Hayma was already affecting outflows from the Miqbaba basin. Right from the outset of the city scheme competition for the dwindling resource between the city and the farmers made aquifer depletion inevitable. Effective complete depletion had occurred by 1991, if not by 1988.

In reality, farmers will not stop irrigating if technology, land and water are available and affordable, and farmers did increase the irrigated area in Al Hayma after 1976. However the abstraction for the city on its own would have exhausted the resource, except it would have taken a little longer if irrigated agriculture had stopped as the original scheme unrealistically envisaged.

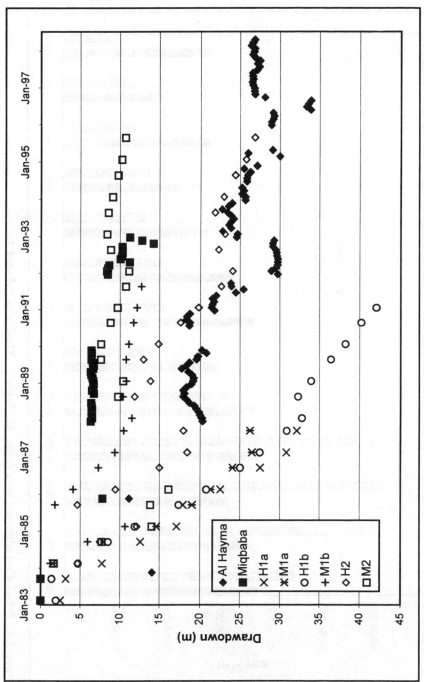

**Figure 2.9 Transient model calibrated hydrographs** Filled symbol=measured, others=modelled, H=Al Hayma, M=Miqbaba

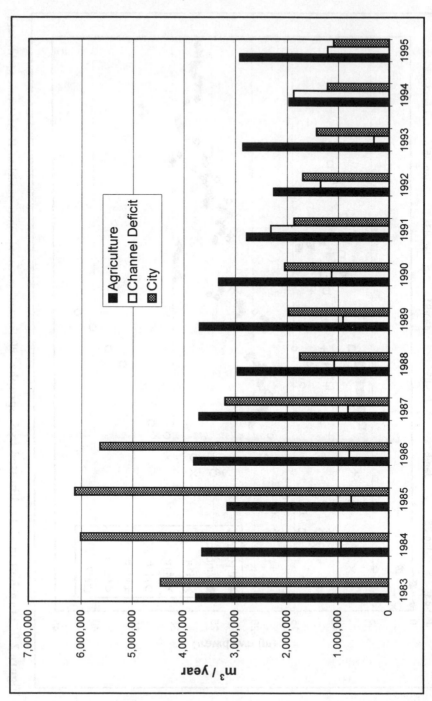

**Figure 2.10  Modelled abstractions from Al Hayma / Miqbaba (Model 2)**

[The "channel deficit" mentioned in Figure 2.10 refers to the excess amount of water which has been drained from the channel areas to the basin downstream, usually in the wet season. If this water had not drained to the basin by the following dry season it could have replaced any irrigation deficit in the channel areas in the dry season. It is considered that during the early 1980s the storage in the channel areas was sufficient to meet this deficit, but that as depletion progressed, the "channel deficit" has approximated the actual deficit. The "channel deficit" thus represents an unknown in the model that could be added to the agricultural consumption of water in the earlier years of the model to get a more accurate assessment of the agricultural consumption. In the later years of the model (when the deficit is larger) it contributes less, if at all, to the estimate of agricultural consumption. In any instance, the overall picture of 'who took the water?' is clear.]

*Aquifer Recovery: the Hydrogeology of Potential Resource Reconstruction*

The establishment of a calibrated mathematical model permits the tentative exploration of future trends. It has been used to predict how long it would take the aquifer to return to its 1976 steady-state condition if groundwater abstraction for the city remained at the (low) 1995 level and if irrigated agriculture ceased. Under these conditions the Miqbaba basin would recover after five years and the Al Hayma basin after ten years (both +/- 20%).

The history of the Al Hayma valley water resource has been examined as a case study which provides a necessary introduction and background to the issues of water management in the wider Ta'iz area. The reason for selecting Al Hayma is that the demise of this particular resource has had the single biggest impact on many of the wider water management issues and a better informed discussion of them in the following chapters is possible.

**Water Resources Pollution**

This section assesses the impact of water use practices in the Upper Wadi Rasyan catchment on water quality. One consequence of falling groundwater levels has been reduced base flows, which in turn has resulted in less dilution of pollutants, that is, higher concentrations in surface and groundwaters.

*Methodology*

A survey of surface water electrical conductivity (EC) was conducted as part of this study during November-December 1995, September and November 1996 and March 1997. Analyses of groundwater and surface water were also obtained from sampling undertaken as part of the NWRA well inventory (1996) and a study by van der Welle (1997) in autumn 1996. The NWRA well inventory analyses (1996) were undertaken in the NWSA laboratory in Ta'iz and the ionic balance was usually worse than +/- 20%. Only where their analyses were better than this, have they been used in compiling Figure 2.11 Although the level of error in the inventory analyses is unacceptable, for many locations it comprises all that is available and their data are consistent with those of van der Welle.

Van der Welle also encountered problems of laboratory analytical reliability and only where the data from her study were considered reliable have they have been included in this discussion. The samples reported in this study were collected at various dates in the hydrological year. However, it should be noted that concentrations generally reach levels around 50% higher at the end of the dry season in February / March than in most of the rest of the year. Sampling was generally not conducted immediately after a rainfall event when concentrations would be appreciably lower than "normal".

*Polluted Areas*

Figure 2.11 indicates that apart from a few upper reaches of streams and Wadi Ad Dabaab the only water courses which flow for more than 6 months per year are heavily polluted. Using the irrigation salinity hazard criteria, Figure 2.11 divides the valleys into those whose groundwater or surface water exhibits a high hazard (750-2250 µS/cm EC) and those with a very high hazard (>2250 µS/cm). This division is even higher than the WHO limit for drinking water of 1500 µS/cm which has been adopted by NWSA. The EC of rainfall in the area has been measured at 20 µS/cm and spring and surface water flows in the mountainous areas are as low as 300 µS/cm. However, all samples from wells and streams in wadi alluvium are above 750 µS/cm. The area of >2250 µS/cm EC is confined to the east to west flowing central zone of the Upper Rasyan catchment.

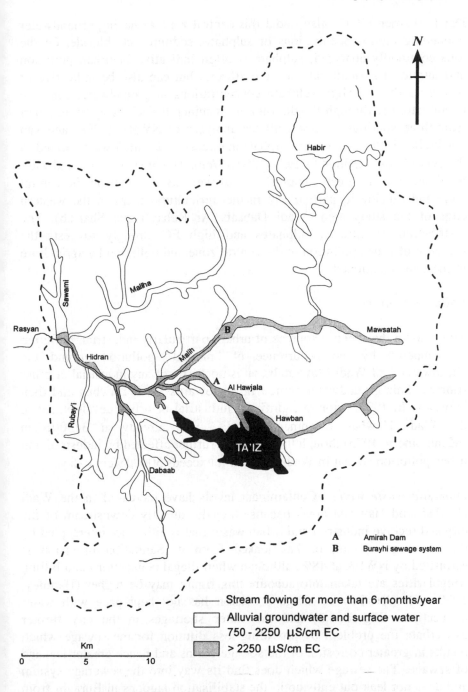

**Figure 2.11  Environmental impact on surface waters of the Upper Wadi Rasyan catchment**

Dar El Yemen (1997) also noted this central zone as having groundwater containing high concentrations of sulphate, sodium and chloride. Of the ions commonly analysed, sulphate is often indicative of urban pollution and nitrate of agricultural use of fertilisers, but can also be indicative of sewage pollution. High sulphate concentrations in groundwater can also occur naturally through the dissolution of minerals such as gypsum. Apart from three spurious analyses, all the analyses of NWRA (1996) and van der Welle (1996) containing sulphate in excess of 250mg/l were located in the central zone defined by EC >2250 µS/cm. The areas with groundwater of nitrate content greater than 22.6 mg/l were located both in the central zone and outside it, particularly in the agricultural areas in the western edge of the study area (Wadi Dabaab, Ar Rubay'i and Shar'ab). The distribution of nitrates, sulphates and high EC strongly suggests the presence of urban pollution in the central zone and pollution by agriculture in the areas mentioned.

*Pollution Sources*

Tying down the specific sources of urban, particularly industrial, pollution is hampered by the occurrence of "natural" pollution around the "headwaters" of Wadi Mawsatah that is upstream of any industrial or other sources. This groundwater source flows towards Wadi Hawban and then Hawgala. In 1974, prior to significant industrial or urban development in the Wadi Hawban catchment the EC was already at 3750µS/cm (Montgomery, 1975) though this has been further affected by industrial and urban pollution so that in Wadi Hawban it exceeds 5000µS/cm today.

*Domestic waste water* Contaminant levels have increased in the Wadi Hawban and Hawgala areas because they lie directly downstream of the city and receive industrial and urban waste that is either not intercepted by the sewerage system, or has leaked from it. Sanitation provision is estimated by NWSA at 48%, although when illegal connections and billing irregularities are taken into account this figure may be higher (Handley, 1999a; 9). Certainly sewerage provision has not kept pace with water connections (Figure 4.3) and the water shortages in the city further exacerbate the problem by providing less dilution for the sewage which results in greater corrosion of the sewage mains and hence greater leakage of sewage. The sewage which does find its way into the sewerage system and does not leak out ends up in the stabilisation lagoons in Burayhi from whence it flows down Wadi Malih. The net result is that the surface water and groundwater downstream of Hawban/Hawgala and Burayhi are polluted by the city effluent. Farmers use this water for irrigation but can only grow millet because of its high salinity resistance and even then they notice a rapid deterioration in yields and soil quality. In the absence of any other sources, the inhabitants of the downstream areas are forced to use this water not only for irrigation but also for domestic purposes (Photo 1) and even drinking in some instances, knowing full well its origin. Waste water flows from the main city sewers have been measured at 10,000m³/day in the dry season.

*Industrial Waste Water*

**Table 2.15  Summary of the major industrial waste discharges in the Ta'iz area**

| Factory | Treatment | Discharge Location | Discharge Method | Waste Type |
|---|---|---|---|---|
| Hawban Factories 1 | Activated Sludge | Wadi Mawsatah | Partially submerged pipe | Food Production Waste |
| Hawban Factories 2 | None | Adjacent to Factories | Lagoons and Cess Pits | Sewage and Plastics Waste |
| Soap and Ghee | New Treatment ? | Hidran | Lagoons | Oils and Fats |
| Paint Factory | None | Adjacent to Factory? | Pit? | Paint Waste |
| Sheibani Food and Drinks | None | Bir Basha | Pipe | Sewage, Food and Drink Production Waste |
| Soft Drinks | None | Wadi Dumay- nah | Wadi bed to Lagoons | Sodium Hydroxide |
| Proctor and Gamble | None | Adjacent to Factory? | Pit? | Not Known |

Table 2.15 summarises major industrial waste sources. The industrial complexes of Hayel Said at Hawban include Nadfood, Genpak and YCIC, and of Sheibani at the western edge of the city, include Food Products and Paradise Juice.

Occasionally, factories, which normally pipe waste away, have been known to dump water in depressions near to the factory when pumping equipment has failed. Heavy metal content (mg/l) of groundwater samples near to three factory discharge points confirms the occurrence of industrial pollution:

**Table 2.16  Heavy metal pollution** (from van der Welle, 1997)

| Parameter | WHO Guide- line | Sample location / pollution source | | |
|---|---|---|---|---|
| | | Wadi Rubay'i / Paint Factory | Wadi Mawsatah / Hawban Factories | Wadi Hidran / Soap and Ghee |
| | 0.05 | 0.22 – 0.14 | 0.13 | 0.19 |
| Lead | 0.05 | 0.14 – 0.13 | 0.34 | 0.37 |
| Cadmium | 0.005 | 0.01 | 0.03 | 0.03 |
| Manganese | 0.05 | 4.7 | | |
| Nickel | 0.05 | | 0.08 | 0.07 |

The well from which the paint factory pollution was detected, and those immediately downstream of it, are used by approximately 1000 people for drinking water collection. Local inhabitants complain about the odours from the discharge point in Wadi Mawsatah. When they complained previously, the factory extended the discharge pipe to its current position. It would appear that the activated sludge treatment of this waste source is inadequately maintained or is insufficient for the demand load. Although some form of waste treatment has apparently been installed at the Soap and Ghee factory since these samples were collected, it is as yet unclear whether the treatment has reduced the level of pollution significantly. Discharge from the Soap and Ghee factory has been to lagoons perched above a small wadi immediately to the north of Ta'iz city dump (Photo 4). Seepage and evaporation approximately keep pace with supply. There is a seepage face at the base of one of the lagoon dams. Concentrated liquid waste from the city is also tankered to the dump. The environmental damage caused by years of subsurface seepage cannot be undone by whatever waste water treatment is undertaken now.

The build up of heavy metals and other toxins in the food chain in the agricultural areas of the Rasyan catchment downstream of the city and the factories has not been investigated.

## Environmental Impact on the Upper Wadi Rasyan Catchment

*Groundwater Levels*

The environmental impact of abstraction levels in the Upper Rasyan catchment have been felt most in Al Hayma. The development of the wellfields in Al Hayma have been felt as far downstream as the sewage lagoons at Burayhi with declining wadi flows and groundwater levels and the resulting drying up wetlands or xazaga. This latter effect has been perceived as a positive one by locals who have been able to increase the area of land available for agriculture. The reduction in wetland has also been accompanied by a local decline in malaria and bilharzia. Groundwater development also affects upstream areas in that agricultural development, in search of dwindling sources of water moves upstream, as the satellite imagery demonstrated in Al Hayma.

Water levels have also notably declined in the Hawban area of the Hayel Said factories, lower Wadi Dabaab adjacent to the Soap and Ghee factory and in the area of Ta'iz city itself.

*Pollution Levels*

Nearly all of the streams in the Upper Wadi Rasyan catchment which flow for at least six months of the year are polluted by, or entirely comprise, domestic and industrial waste water. Groundwater pollution extends even further (Figure 2.11). Domestic waste water accounts for most of the total pollutant load by virtue of the large volume of untreated waste, a significant proportion of unsewered properties and leaking sewers. Some industrial waste water is known to contain heavy metals but levels are not monitored.

*Conclusions*

Falkenmark and Lundqvist's (1995) correlation of the regions with the most rapid population growth and stagnating food production, coinciding with extreme vulnerability due to hydroclimatic constraints, aptly describes

Ta'iz. That vulnerability is expressed in declining water availability and quality.

*Water quantity* The Al Hayma aquifer took approximately four years to deplete and will take nine years to recover if no more irrigation takes place and the city abstracts current minimal quantities. The next chapter attempts to put a value on the loss of agricultural production and livelihoods if resource degradation were to be reversed.

*Water quality* The full cost both now and to future generations of the pollution in the Ta'iz area is, perhaps, impossible to estimate (Allan and Karshenas, 1996; 125). The loss of soil fertility, the increase in toxin levels and especially heavy metals in the food chain and the legacy of polluted groundwater and surface waters of the area are severe. The impacts of pollution on surface and groundwaters was immediate, and the prolongation of polluting activity over the past twenty-five years has ensured a significant build-up of pollutants in the unsaturated and saturated zones. Although some of these environmental impacts may be reversible, many of them are not (Karshenas, 1992; 2, Feitelson and Haddad, 1998, Pearce, 1993; 50). Others would take a long time to rectify and need a lot of fresh water to dilute the impacts to acceptable levels. That water is simply not available.

As with many Yemeni urban areas, an environmental "time-bomb" has been set off which will leave its mark on generations to come. Even the "reversible impacts" may, in reality, be irreversible due to the deficit of economic capacity and political will. The socio-economic and socio-political aspects that have contributed to this environmental disaster and which might contribute to any resource renewal are considered in chapters three and four respectively.

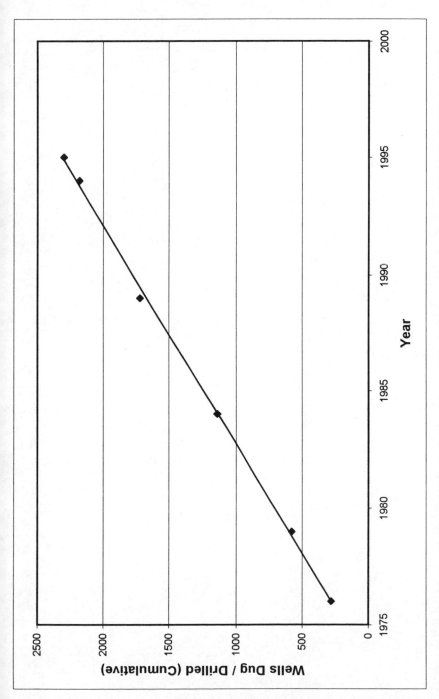

**Figure 2.12 Number of wells drilled and dug, Upper Wadi Rasyan catchment, 1976-1996 (NWRA, 1996)**

# 3 A Socio-Economic Map of Water Allocation and Use

The depletion of the water resources of Al Hayma analysed in the previous chapter exhibits in microcosm the national picture. The annual renewable water resources of Yemen have been estimated at $2.1Bm^3$, or $125m^3$/capita/year. This is equivalent to about 10% of the regional MENA average and compares with a world average of $7500m^3$/capita/year (World Bank, 1997; Davies and Sahooly, 1996). The reason is not so much climatic (the area has a higher proportion of rainfall than the regional average) as demographic. By the year 2025 it is estimated that population growth will result in renewable water resources per capita of 72 $m^3$/capita/yr (ibid.). As a reminder, this compares with individual human water needs of $1m^3$/yr for drinking, $100m^3$/yr for domestic purposes and $1000m^3$/yr to feed her.

In the context of such shortage, it is suggested that the survival of Yemen during the next century depends, not so much on how much water is available, but on how well the scarce water resources are used. This purpose of this chapter is to:

1. evaluate the relative returns to water of agriculture and industry in economic and livelihood terms,
2. describe the socio-economic impacts of the water scarcity and pollution problems described in the previous chapter particularly in the urban area, and
3. demonstrate the adaptive capacity of households and water-related businesses to the scarcity in the context of equity.

## Agricultural Sector Water Use

*National and Governorate Context*

The most impressive feature of the Yemeni landscape is the terracing. In North Yemen this agricultural system dates back to at least 1000BC when, it is thought, 300,000 people derived livelihoods from 20,000 hectares of terraced land. Yemen is one of the few places on earth where rainwater harvesting and runoff agriculture has been practised continuously since earliest settlement (Prinz, 1994).

The shift from rainfed and runoff-fed agriculture to irrigation, particularly groundwater based irrigation indicated in Table 3.1 is very recent, and was part financed by remittances earned by up to one million Yemenis working in Saudi Arabia and the Gulf from the mid-seventies up to the Gulf War in 1990. With so many people absent from their home villages, the same period witnessed a sharp decline in the proportion of labour available for agriculture. Some authors suggest this was the cause of degradation in the traditional rainwater harvesting infrastructure (Varisco, 1991).

Today agriculture occupies about 5% of the land in Yemen, though there has been some reduction in cultivated area since 1970, the loss being from rainfed arable land (Bamatraf, 1987; 595). Table 3.1 indicates the decreasing agricultural proportion of GDP and the decreasing proportion of the labour force and of the population involved in agriculture. Because of the underlying population growth, however, the rural population dependent on agriculture has doubled from 5.8 million in 1970 to 11.5 million in 1990 (World Bank, 1993).

Despite the increasing availability of labour to work decreasing areas of land, and despite increasing mechanisation and use of fertilisers and pesticides, yields have declined or at best remained static. Growth areas in agriculture have been confined to fruit, vegetables and qat production. These comprise the main crops that use groundwater and the increase in their production has been the major factor contributing to unsustainable groundwater consumption. The huge growth in the irrigated areas (250% growth between 1970 and 1996) has been outstripped by the population growth so that the irrigated area per Yemeni is now about 17.5m by 17.5m. The Ta'iz governorate accounts for about 10% of the total cultivated land in Yemen but 16% of the population.

## Table 3.1 Yemen national agricultural statistics

| | Units | 1970 | 1975 | 1980 | 1985 | 1990 | 1996 |
|---|---|---|---|---|---|---|---|
| YR:$US | ratio | | | | 6 | 8 | 100 |
| % of Labour in Agriculture | % | 71 | 66 | 62 | 59 | 58 | 58 |
| % of Population in Agriculture | % | 78 | 71 | 67 | 65 | 63 | |
| Agricultural share of GDP | % | 45 | | | 20 | 17.8 | 15 |
| **Cultivated Land Area** | | | | | | | |
| Cultivated Area | M ha | 1.3 | | | | 0.9* | 1.1 |
| Irrigated Area | M ha | 0.21 | | | | | 0.49 |
| % irrigated area | % | 16 | | 17 | | | 45 |
| Cereal | 000 ha | 1080 | | 850 | 860 | 850 | 700 |
| Fruit and Vegetables | 000 ha | 39 | | 75 | 69 | 92 | 136 |
| Qat | 000 ha | 8 | | 70 | 75 | 80 | 91 |
| **Production** | | | | | | | |
| Cereal | 000 T | 845 | | | | 491* | 731 |
| Fruit and Vegetables | 000 T | 20 | 73 | 595 | 835 | 1007 | 1094 |
| Qat | Million Bundles | 35.2 | 154 | 352 | 387 | 501 | 592 |
| **Yields** | | | | | | | |
| Maize | Kg/ha | 2250 | 2140 | 1510 | 1149 | 1274 | 1240 |
| Sorghum/Millet | Kg/ha | 795 | 800 | 888 | 416 | 764 | 800 |
| Potatoes | Kg/ha | | | 12364 | | 11830 | 12840 |
| Tomatoes | Kg/ha | | | | | 15528 | 15210 |

*1991

*Sources*: Bamatraf (1987), World Bank (1993 and 1998), Statistical Yearbook (1996).

*Methodology of the Socio-Economic Surveys of Rural Water Use
undertaken in 1995 and 1996*

Surveys were designed to obtain the data that would provide an understanding of water use in descriptive terms and also permit an evaluation of returns to water. Mainly due to limitations in transport availability, it was necessary to employ an essentially "rapid rural appraisal" (RRA) method in examining water use in agriculture (Chambers, 1992, Cromwell, 1990; 23, Pretty et al, 1989; 54-62, Chambers and Carruthers, 1986). This was supplemented by participatory (PRA) methods (ibid.), mapping of wadi areas and satellite imagery analysis described in chapter 2.

The field work was conducted within the context of, and with the financial and logistics support of, two UNDP surveys, in November / December 1995 and September / December 1996. More detailed accounts of the surveys and their findings are reported in Handley (1996a) and Dar El-Yemen (1997) respectively.

*1995 survey* Aspects of the survey discussed within this section include:

1.      The overall physical, demographic*, and occupational*, aspects of five villages.
2.      Land ownership and tenure*.
3.      Water availability and use from surface and groundwater sources.

*Although the methodology was determined and the survey carried out by the author, the development of the methodology for the RRA part of the survey included valuable sociological inputs regarding these aspects from Cecile de Rouville, international consultant attached to the UNDP project, and are gratefully acknowledged. The survey obtained other information than that directly relevant to agriculture, some of which is used in other sections.

The five villages were selected to give a representative agro-ecological coverage of the governorate excluding the Tihamah. Their locations and names, together with those of other nearby villages included in the survey are shown in Figure 3.1. They all comprised less than 500 people. Criteria determining the agro-ecology were selected as rainfall, soil type and irrigation method. Within that context areas which had supplied Ta'iz with water, or which are being considered for future supply to the city, were included. The villages chosen varied in rainfall between around 300 and 700mm/yr.

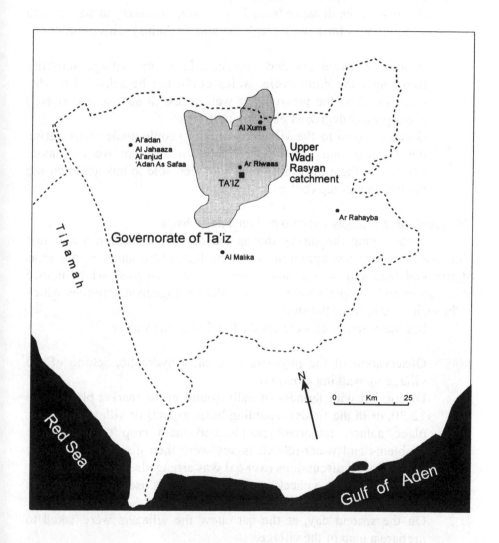

**Figure 3.1 Locations of villages included in the 1995 rural survey**

Although the USAID satellite imagery-based soil map (1983) indicated a variety of soils, in the field the soils of the cultivated parts were quite similar, comprising immature, deep light loamy loess soils with horizons of pebbles and cobbles and occasional clay layers. Adjacent uncultivated runoff areas were rocky with little to no soil and wadi bottom areas comprised fluvially deposited sands and gravels.

Because of the distance from Ta'iz, it was necessary to stay in each of two of the villages for a week (Al 'adan and Al Xums). This permitted:

1.    A greater awareness and appreciation of the village activities throughout the daily cycle, which could not be achieved by day visits (such as the pressure for well access at sunrise and sunset) and a greater degree of observer participation.
2.    Greater access to the villagers – many people made visits during the evenings and often talked about issues that were of more concern to them than those they had been able to talk about during the day in a group context.

The other three villages were visited on a daily basis.

Undertaking the survey during the dry winter season meant that although there was less agricultural activity than in the summer, the areas of irrigated land and rainfed land were easy to distinguish which helped observation and comprehension of the water management activities which is the main emphasis of the study.

Six main activities were conducted during the survey:

1.    Observation of the physical, especially hydraulic, setting of the village by walking around it.
2.    Discussion with farmers (usually found in the market place up to 12:30, or in the fields) regarding basic aspects of village structure, place names, important people and main crop types. Village problems and water-related issues were then discussed. A venue for afternoon discussions over qat was arranged.
3.    At the qat chews a checklist of questions was used covering all the aspects of the survey.
4.    On the second day, at the qat chew the villagers were asked to prepare a map of the village.
5.    Short impromptu excursions were arranged by the villagers to places that they considered hydraulically important.

6.      In the evenings individual villagers came to ask questions and clarify or correct comments they or others had made during the day.

The following points were noted regarding the methodology:

1.      Although interview sampling would have been much more straight forward if the desired interviewees (farmers) could have been isolated, in Arab society opinions are often public property and they are to be developed by consensus and expressed by those considered worthy. Market place and qat chew discussions were dominated by one or two spokesmen but with perhaps 20 men listening and discussing what was being said amongst themselves and often the speaker was corrected on points. Discussions were often heated and dominated by the visibly active males (typically between 20 and 50 years old). When the local shayx was absent the role of spokesman was taken by an influential farmer / well owner.

2.      Another bias came with the "technical" questions involving measurements and numbers generally. Although efforts were made to simplify them as much as possible using various RRA techniques, these questions were really only understood by those between 15 and 30 years old and sometimes crude estimates had to suffice. On occasions it appeared that numerical answers were purposely left as vague as possible, perhaps from fear that the survey was part of a government tax assessment or plan to take their water. Fear of the shayx appeared to hold people back from answering. Because the survey was linked to the UNDP, questions could be answered in a manner more likely to attract foreign aid.

3.      During the visits to hydraulic locations considered important to the villagers it was possible to visit homes on a more individual basis which permitted a refinement and, on occasions, a complete reversal of previous comments.

4.      In preparing the village map the local school teacher was usually chosen as the cartographer. Other men would direct what he drew and teenage boys often had the best grasp of the details. In nearly all the villages, and without prompting, the map focussed on the wells, the routes of spate flows / streams and the alignment of fields. The shapes and sizes of features were less important.

5.      The absence of a woman interviewer prohibited the interviewing of women, although it was sometimes possible to talk to older women at the well head.

*1996 survey*  This survey comprised mapping the land use of the Upper Wadi Rasyan catchment in September and December 1996 to field check the extent of rainfed and irrigated agriculture noted on the satellite imagery. A brief field visit to key areas was also made in spring 1997 to check the extent of third season irrigation. Because of the high intensity of cultivation in wadi areas indicated on the satellite imagery, the wadis were surveyed in detail to determine crops grown and irrigation methods. Farmers were also interviewed regarding aspects of farm inputs and outputs to permit a rudimentary assessment of farm budgets.

## Survey Results

*Land ownership and tenure*  This field of study does not contribute directly to the evaluation of returns to water or the description of water use practices, however it is included here because it forms relevant background regarding agricultural water users. Table 3.2 summarises the main findings regarding rent, zakaat and tenancy agreements.

The non-contiguous distribution of owned land and extreme fragmentation of land made an evaluation of land ownership and tenure very tenuous (a problem also noted by others, Mitchell and Escher, 1978). Average holding sizes of 1.2ha with 4.6 parcels per holding were noted for the Ta'iz governorate (Bamatraf, 1987). NWRA (1999a) also notes that the majority of farms are under 1ha in size. Individual plots (haql, hawd or faddaan) are demarcated (tahdiid). The non-dividing up of fields upon inheritance leads to confusion of ownership but permits easier farming. Lack of land division may also be putting off the fateful day of deciding who owns what. Resolving the latter occupies most of the time of the shuyuux (plural of shayx), uses up most of the bullets that are fired in Yemen in earnest, and is further complicated by daughter's inheritances which should be half that of the sons but is often withheld by force. Since the Gulf War, migration to the cities does not result in fewer people to farm the land because rural population growth still outstrips migration. However, lack of work on the land appears to lead to a higher rate of migration. There is little incentive to measure land except for the purpose of sale. However sale is quite rare. Selling land was described as similar to the ultimate dishonour of "exposing one's women folk". Remittances from Saudi Arabia and the Gulf were spent on houses, vehicles, wells/pumps and businesses, but not on land purchase. Only under times of great financial stress do people sell land, as occurred in Al Jahaaza during hard times under the reign of 'imaam 'ahmad.

The average proportions of farmer-owned and sharecropped land in Ta'iz – Ibb (World Bank, 1986; Annex I) were 84.1% and 4.5% with mixed holdings accounting for 11.4%. This compares with 60.2% owner-farmers and 13.2% pure tenants in the Upper Rasyan catchment (NWRA, 1999a). The average for North Yemen (Bamatraf, 1987; 581) is 77.4% owned, 3.5% sharecropped, 0.3% waqf and 18.8% mixed ownership. NWRA (1999a; 17) suggest a greater proportion of sharecropped land and fixed-rent land is irrigated than is the case with owner-farmed land.

## Table 3.2 Rent, zakaat and tenancy agreements

| | Sharecrop Rent | | | Zakaat | | Who Pays? | Length of Agreement |
|---|---|---|---|---|---|---|---|
| | Irrigated | | Rainfed | Irrigated | Rainfed | | |
| | Land Rent | Water Rent | Land Rent | | | | |
| Al 'adan | 1/4 | 1/3-1/4 | 0-1/5 | 3-5% | 3-5% | | Semi-permanent |
| Al Jahaaza | 1/3 | 1/3-1/4 | | | | tenant | |
| Al 'anjud | | | | | | | Semi-permanent |
| 'adan As Safaa | 1/3 | 1/3 | | 3% | | owner | >5yrs |
| Al Xums | 1/4 | 1/4 | | 2.5% | 10% | owner | 1-20yrs (av.5yrs) |
| Ar Rahayba | 1/4 | 1/4 | 1/2 | 5% | 10% | tenant | 5-12 yrs |
| Ar Riwaas | 1/4 | 1/4 | 1/2 | 3% | | owner | 20 yrs |
| Al Malika | 1/4 | hourly | | 100-200 YR | 100-200YR | tenant | 1 season |
| Al Malika (waqf) | 1/4 | hourly | | 10% | 10% | tenant | hereditary |

In the villages included in the 1995 survey very little hired labour is used, although some landowners employ help during harvest and wealthier landowners hire labour for ploughing, sowing, harvest, irrigating and tree trimming, with some employing labourers on a full time basis. Labourers may sharecrop and own small portions of land. Most farmers repair terraces themselves but skilled labour, particularly ploughmen, are hired for their skill and equipment. Payment for hire may be in crops, though it may be in cash from the few who have access to cash through the sale of water or qat.

The zakaat (tax) is supposed to be 10% for rainfed production and 5% for irrigated crops (Mitchell and Escher, 1978; 40) and paid to God. In reality, 3-5% is paid to the government on irrigated or non-irrigated crops via the local leadership, although both the reallocation and reduction are seldom remarked upon or even noted in the literature. Cases of enforced payment of zakaat are reported (Handley, 1996a; 17) and the system is not free from irregularities.

Rent is paid as a proportion of the crop by the sharecropper to the landowner. It seems to make no difference whether the rent is paid in cash or crop in the case of grains, but rent for cash crop land and water rent (particularly for qat) are paid in cash. There was no indication of different rents being paid for different crops. Waqf rental payments used to be paid to the mosque or designated charity (Mitchell and Escher, 1978; 95) but it now appears to be paid to the state. Water rent in Al 'adan and Al Jahaaza is reduced from one third to one quarter if the land was gayl (stream) fed rather than supplied from underground sources. Sharecroppers sign rental agreements in the presence of two witnesses and the 'amiin (local officiary), who stamps the document. If the sharecropper stops paying then the contract is terminated. Although similar rental figures to those given in Table 3.2 are widely quoted, some frank conversations suggested that, in reality, the rainfed land sharecropper pays as little as possible. If the crop is poor he gives a token amount and if good, and depending on how tough the landowner is, he may give as much as 20%.

Sharecroppers have to obtain permission from the landowner to dig a well. The owner cannot take the well from the renter who dug it if permission was obtained. The landowner can only buy the well from the renter if the latter agrees. If the landowner wishes to sell the land and move the sharecropper off then he has to pay for the digging of the well. Similar rules apply to planting trees and repairing terraces.

*Irrigation practices*   Because of the seasonal rainfall variation, rainfed cultivation (daahi) is restricted to one summer season. With irrigation, two, or even three cultivation seasons are possible. Because the rainfall events are intense and of short duration, rainwater harvesting of the spate (sayl) has to be self-regulating and adequately robust. Diversion of the spate from a feeder wadi bed (saa'ila) to the fields is via a diversion structure (masqa, pl. massaaqi). If the sayl is too strong there may be destruction of the massaqi and erosion (injiraaf) of the land. The masqa is constructed to reduce the erosive power of the sayl as well as to obtain the water, and it is reconstructed if it is destroyed. There are two types of masqa, one is the runoff routes from the mountains to the main wadi, the rights to which are ancient and are attached to the land. On a smaller scale, in the main wadi floor, the spate water derived via the masqa from the main wadi bed appeared to be the right of whoever could manage to divert the spate flow.

The other gravity-fed form of irrigation is the spring-fed stream (gayl). These may be seasonal or perennial, usually depending on the degree of exploitation upstream. In some instances, the spring source was controlled by the solid geology such as at Al 'unjud and Al 'adan where the Tawilah Sandstone outcrops with artesian flow (Photo 3). The gayl is diverted in open channels to fields in fixed rotation referred to as "al 'awwal bil 'awwal" or " al 'ala bil 'ala", that is, "highest first".

The valley alluvium constitutes the most important source of water for pumped irrigation. Exploitation begins with hand dug wells which are successively deepened as the resource declines until the rockhead is reached. As water levels decline further the well may be extended by up to 20m using pneumatic tools. After that those wealthy enough have to resort to drilling. Dug wells are typically 1 to 2m in diameter and are reinforced-concrete lined using corrugated iron or plywood formwork. Construction proceeds downwards, concreting about 1-2m at a time. Caving of the unsupported sides during construction may result in ground surface subsidence of up to 2-3m around the shaft. Wells are dug in winter to ensure maximum depth which is typically 5m below the summer water level. Dug wells may reach up to 70m deep. Pumping from permanent dug wells in wadi floors is usually by shaft pumps belt driven by 24 to 36 HP diesel engines. The pumps are connected to 3-inch or sometimes 4-inch steel pipe distribution networks via gate valves. More temporary arrangements may be used, especially if the groundwater level is not very deep, utilising 2-inch petrol driven suction pumps lifting water to a maximum elevation of 30m above the pump from a maximum depth of 5m below pump suction. The rising main and distribution is by 2-inch flexible hose. The wells may be abandoned and re-dug after spates.

Wells drilled by farmers are usually between 100 and 300m deep and, like the dug wells, are deepened as water levels decline. They are usually finished in 8-inch casing and are open hole below the water surface, although torch-slotted casing is sometimes used. As with the permanent dug wells, shaft driven pumps and piped distribution systems or open channels are used. NWRA (1999a) note that the preference for open channel or piped systems is largely determined by whether the topography permits the construction of channels rather than other factors. With piped systems a flexible hose is typically used at the field to direct flow more easily, particularly in larger fields.

Table 3.3 gives a summary of the different hydraulic regimes which are used for irrigation and the different regimes in which the regimes are found. Working down the table, the hydraulic regimes are cumulative, the land irrigated by a seasonal stream, for instance, will also receive water from the spate and be rainfed. The villages can be considered to be at various stages in a process of exhausting the aquifers. Large differences in wealth are attached to being able to cultivate three seasons of cash crops as opposed to only one season of subsistence crops.

As the water level falls in an alluvial aquifer penetrated by shallow dug wells, the wells dry up. In this situation, the only way a farmer can maintain cultivation for two or three seasons is to drill a deeper well into the underlying rock. The costs and potential for conflict of doing this may deter him. It was observed that conflict over water access is concentrated in this situation where the water level had reached the base of the alluvium. By contrast, both when groundwater resources are plentiful, and when they are insufficient for all but the few farmers who were able to drill deeper wells, the potential for conflict decreases. Uphoff et al (1990; 27) expected user participation in water resources management to be most readily forthcoming in the middle range of resource availability, and Thompson thought a 'bounded egalitarian group' most likely to emerge where there is resource depletion (1988; 67). However, the Yemeni experience is the opposite, with resource depletion producing the greatest conflict potential.

Whether a farm has a surplus of water relative to land or land relative to water availability can be directly correlated with farm practices regarding cropping, the extent of supplementary irrigation and feasibility to sell water (Handley, 1996b; 11). By their cropping practices farmers clearly demonstrate their aim to achieve a balanced use of land and water without underutilising one or the other. A surplus of water is rarely the problem since there are almost bound to be some surrounding farmers with water deficits to sell to. A surplus of land is usually the case. This underscores the principle that greater productive efficiency of water use through improved irrigation techniques, for example, will only result in use of more land and will not in itself reduce groundwater withdrawals.

Ploughing is mostly by oxen, and the furrows (tilm) are spaced according to crop, for example cereals 40cm and potatoes 50cm, although there is considerable variation in this. Around 10 tilm make a sabba. The sabba is formed by raising the soil higher than the furrow crests so that an enclosed irrigation basin is formed. In some areas the sabba is called a tilm and comprises 4 to 6 furrows. The individual sabba is filled with water to the height of the furrows (10-15cm) and the water is then switched to the next sabba. The fields are irrigated from upstream downwards.

**Table 3.3  Irrigation regime and conflict potential**

| Hydraulic Regime | Environment | State of Alluvial Aquifer | Technology | Aquifer* Exploited | Village (Cultivated Area) | Number of Seasons | Economy | Conflict | Environmental Impact |
|---|---|---|---|---|---|---|---|---|---|
| Runoff area | Mountains | | | | Most Dwellings | 0 | | | |
| Rainfed only | Middle and Lower Catchments of Broad Wadis and Small Catchments of Narrow Wadis | Exhausted | | None | Miqbaaba | 1 | Minimised | Least Vulnerable | (Most Affected) |
| | | | | None | Al 'adan, Al Jahaaza | 1 | | | |
| | | Almost Exhausted | Drilled Wells | Volcanics | Al Xums - mid wadi | 1-2 | | | |
| | | | | V & A | 'adan As Safaa | 1-2 | | | |
| | | | | V & A | Al Malika - mid wadi | 1-2 | Maximised | | |
| Sayl - Runon area | Upper Broad Wadis and Wadi Margins | | | V & A | Al Xums - Wadi Edge | 1-2 | | | |
| | | | | V & A | Al Malika - Wadi Edge | 1-2 | | | |
| Gayl - seasonal | Large Catchments with slightly depleted to un-depleted upper reaches | Sufficient | Dug Wells | Alluvium | Ar Riwaas | 2-3 | Minimised | | |
| | | | | Alluvium | Ar Rahayba | 2-3 | | | |
| Gayl - permanent | Large Catchments with undepleted upper reaches and/or Geological Control | Almost Unaffected to Too Much Water | | A & T | Al 'anjud | 3 | Minimised | Most Vulnerable | (Least Affected) |
| | | | | A & T? | Al Hayma pre-project | 3 | | | |

* Aquifers:

V & A    Volcanics & Alluvium

A & T    Alluvium & Tawila Sandstone

Kuraath (everlasting onion) is grown in a separate rectangular tilm without furrowing. Although it was consistently stated that women played no role in irrigation activities, women, including axdaam (servant caste) were occasionally observed controlling the flow of water into fields.

Irrigation efficiencies of 35% are reported in an area of open channel irrigation in the At Turba district (World Bank,1986;Annex I;13). Total (well to field) irrigation efficiencies of 72% and 60% for piped and open channel systems in the study area have been estimated (Zagni, 1996). No volumetric measurements of water used in irrigation are made, all sales are by the hour and flow rates are always mentioned in terms of pipe diameters only, 2-inch, 3-inch etc.

*Water use* A summary of the major crops grown, the number of harvests and cropping areas in the field study area in both the major wadis and the non-wadi areas for the Upper Wadi Rasyan catchment is given in Table 3.4 to 2.5ha and 0.5 km$^2$ accuracy respectively. It was compiled from a combination of satellite imagery interpretation as described in chapter 2 and field observation conducted as part of the 1996 survey. The crop types are estimated from field checking and the areas of major crop types may be in error by up to +/- 10%. The calculation of areas of summer grains and qat assume observed approximate intercropping ratios of 1:2 respectively in the lowland areas and 2:1 in the highland areas.

Theoretical water use by agriculture has been calculated as follows:

1. The rainfall distribution pattern (Figure 2.2) has been applied to the cropped areas in 50mm/yr gradations.
2. Cultivation intensities were applied according to field observations of the extent of fallow, unused land, roads and tracks within the areas identified by the satellite imagery thresholding method (see chapter 2). An error of +/- 10% is estimated.
3. The effective rainfall on fields during the period crops are growing is calculated as: Effective Rainfall = Rainfall x Rainfall Efficiency x Seasonal Proportion of Annual Rainfall x Cultivation Area x Cultivation Intensity.
4. Field Irrigation Requirement is calculated as: (Net Irrigation Requirement-Effective Rainfall) x Field Application Efficiency.

## Table 3.4  Agricultual land use in the Upper Wadi Rasyan catchment

### a) Areas of Intensive Land Use in Major Wadis

Cultivated Estimated Summer Crop Areas[b] (ha)

| Summer[a] | Qat | Sorghum or Millet | SI Sorghum or Millet | Maize | SI Maize | Fruit | Vegetables |
|---|---|---|---|---|---|---|---|
| Total 2735 | 182.5 | 1152.5 | 322.5 | 182.5 | 810.5 | 47 | 37.5 |

Cultivated Estimated Winter Crop Areas[b] (ha)

| Winter[a] | Qat | Maize | Sorghum or Millet | Potato | Fruit | Vegetables |
|---|---|---|---|---|---|---|
| Total 1352.5 | 182.5 | 562.5 | 160.5 | 292.5 | 47 | 107.5 |

Cultivated Estimated Third Season Crop Areas[b] (ha)

| Third Season[b] | Qat | Maize | Sorghum or Millet | Fruit | Vegetables |
|---|---|---|---|---|---|
| Total 609.5 | 182.5 | 77.5 | 195 | 47 | 107.5 |

### b) Excluding Major Wadis[a]

| | Land Use Summer km$^2$ | Land Use Winter km$^2$ |
|---|---|---|
| Uncultivated[c] | 486.0 | 789.7 |
| Rainfed Summer Poor Quality Grains | 191.8 | 0.0 |
| Rainfed Summer Barley | 6.0 | 0.0 |
| Rainfed Summer Cultivation Only | 8.9 | 0.0 |
| Small Scale Rainwater Harvesting | 55.0 | 0.0 |
| Qat growing areas irrigated by tanker | 18.0 | 18.0 |
| Intensive Summer Cultivation | 37.5 | 0.0 |
| Irrigated Winter Crops and SI Summer Grains | 4.5 | 0.0 |
| Highland Qat | 70.9 | 70.9 |
| Major Urban Areas | 23.5 | 23.5 |
| Total | 902.1 | 902.1 |

[a] Areas determined from satellite imagery.

[b] Crop area estimates based on field checking during Sept-Dec 1996 and March 1997.

[c] Determined from D.O.S. Maps 1981.

SI Supplementary Irrigated.

The values used for Rainfall Efficiency, Net Irrigation Requirement and Field Application Efficiency were taken from Zagni (1996, who used CROPWAT 7, described in Smith, 1992) for Wadi Warazan and Dhi Sufal, although the rainfall efficiencies may be rather high.

In calculating the pumping requirements for well irrigation a sensitivity analysis was carried out using:

1.      Runon-Rainfall ratios of 0 to 35% (after Eger, 1986; 102-105 and Zagni, 1996; 15). [This ratio is based on the runon relative to the rainfall falling directly on the cropped areas, not the total area]. From the observation of runon-rainfall ratios described in chapter 2, ratios of 0.1 to 0.2 are considered more likely.

2.      Irrigation efficiencies of 0.9 to 0.6. Based on the following discussion, the average irrigation efficiency in the Upper Wadi Rasyan catchment is thought to fall within this range:

> Although average well head to field conveyance distances are quite large (287m from dug wells and 457m from boreholes, NWRA, 1999a; 37), on average 85% of this distance is piped (ibid.). Doorenbos and Pruitt (1977; Table 44) define conveyance efficiency in terms of supply from a main (open) canal which has no real equivalent in the Ta'iz instance. Field ditch efficiency equates more closely to the Ta'iz situation. With the given proportion of piped to unlined delivery systems (85:15) the field ditch efficiency is likely to fall between 0.9 and 0.6 (ibid. and NWRA, 1999a; 38), Table 3.5.

As a crosscheck of the amounts of water required from pumping a comparison with well abstractions can be made. The NWRA well inventory (1996) for the area indicates there are 1310 wells in the area (1073 dug and 237 drilled) which are not dry or only accessed by bucket. Applying mean measured discharges and operation periods gives a total abstraction of 36.9 Mm$^3$/yr. If only the wells actually operating when visited are considered the total abstraction would be 24.6 Mm$^3$/yr. A sensible estimate lies somewhere between these two values and gives reasonable agreement with the values in Table 3.5.

**Table 3.5 Agricultural water requirements (Mm³/yr) of the wadi areas of the Upper Wadi Rasyan catchment** (areas given in Table 3.4)

Major Wadis: Effective Rainfall = 6.89
Field Irrigation Requirement = 24.04

| | Pumping Requirement | | |
|---|---|---|---|
| Field Ditch Efficiency: | 0.6 | 0.75 | 0.9 |
| Runon : Rainfall Ratio | | | |
| 0 | 40.7 | 32.05 | 26.71 |
| 0.05 | 39.27 | 31.42 | 26.18 |
| 0.1 | 38.47 | 30.78 | 25.65 |
| 0.2 | 36.88 | 29.50 | 24.59 |
| 0.35 | 34.49 | 27.59 | 22.99 |

Irrigation from stream flow is particularly important in some wadis. The irrigation requirement from the stream has been calculated on a pro-rata basis from the proportion of wadi cultivation served by the stream relative to the total cultivated wadi area (Table 3.6) using the satellite imagery to determine areas of seasonal irrigation. The irrigation requirement compares reasonably with the measured wadi flows, that is, wadi flows balance the requirement of the areas served. An irrigation efficiency for the stream of 0.6 and a runon:rainfall ratio of 20% have been assumed. The maximum annual abstraction assumes the irrigation requirement is fully met by the stream in the areas using stream water. The minimum assumes the lowest flow recorded is available. Likely water use from streams is thought to approximate the maximum since spates have not been taken into account.

The excess flow at the end of the rainy season provides the outflow of Wadis Malih / Hidran and Miliha from the area, to Wadi Rasyan.

**Table 3.6 Stream irrigated land in the Upper Wadi Rasyan catchment**

| Stream Cultivated Area | | Irrigation Requirement | | Wadi Flows | | | | Annual Abstraction from Streams | |
|---|---|---|---|---|---|---|---|---|---|
| S | W | S | W | S | W | Mini-mum | End Rainy Season | Maxi-mum | Mini-mum |
| ha | | Mm³/season | | lit/sec | | lit/sec | | Mm³/season | |
| 223 | 173 | 1.85 | 1.44 | 215 | 168 | 193 | 393 | 3.31 | 1.71 |

S=Summer, W=Winter

Crop water requirements in major wadis are essentially met by groundwater abstractions and saaqiya irrigation. In the rest of the area this requirement is not met. Although runon alleviates the deficit somewhat, evidence of deficiency is seen in the low quality of sorghum and millet from the rainfed areas which incur an estimated 93% of the total deficit. Often, these crops are only adequate for animal fodder, a grain head not having formed.

In the non-wadi areas the effective rainfall is 65 $Mm^3$/yr leaving a field irrigation requirement of 112 $Mm^3$/yr. If the runon:rainfall ratio in the catchments supplying the non-wadi areas is 0.2 then runon would provide another 31.5 $Mm^3$/yr, totalling around 100 $Mm^3$/yr consumption by rainfed agriculture.

Based on the above discussion, total water use in the Upper Wadi Rasyan area is estimated as comprising:

| | |
|---|---|
| Rainfed Agriculture | 100 $Mm^3$/yr |
| Groundwater Irrigation | 30 $Mm^3$/yr |
| Stream-fed Agriculture | 3 $Mm^3$/yr |
| Human Use | 2.5 $Mm^3$/yr |
| Livestock* | 0.3 to 0.4 $Mm^3$/yr |

The estimated total is around 135 $Mm^3$/yr (+/- 20%).

(*Dar El Yemen, 1997; 145 and Al-Dubby and Taher, 1998; 8.)

On the basis of the number of tankers and wells observed supplying water for tanker-irrigated winter crops (NWRA well inventory, 1996) this use cannot account for more than 0.3 $Mm^3$/yr +/- 50%. This is equivalent to an estimated 3 percent of the total possible demand if the cultivation of only one winter crop is assumed. An amount of water not exceeding this may also be pumped from boreholes in wadi bed areas to highland qat farms.

*Returns to Water*

In examining agricultural returns to water the NWRA (1999a) data have been used. The NWRA study comprised a survey of 21 villages to examine socio-economic aspects of agriculture and irrigation in the Upper Wadi Rasyan Catchment. Crop yields quoted in the literature (Table 3.7) are somewhat contradictory although the NWRA data are the most recent, local and from the largest sample. Yields in Ta'iz approximately reflect the national average but suggest scope for the use of improved varieties.

**Table 3.7 Crop yields** (Tonnes/ha)

| Area | NWRA (1999a) Draft Ta'iz | Statistical Yearbook (1996) National | Mitchell &Escher (1978) Ta'iz | Bamatraf (1987) National | El-Lakany (1978) (National) Ta'iz |
|---|---|---|---|---|---|
| SI Sorghum Grain | 1.1 | | | | |
| RF Sorghum Grain | 1.1 | 0.8 | | 2.8 | (0.8) 0.9 |
| SI Maize Grain | 1.3 | 1.2 | 4.4 | | |
| Stalks | | | 8.8 | | |
| RF Maize Grain | 1.3 | | 3.6 | 2.5 | (1.1) 1.0 |
| Stalks | | | 8.0 | | |
| RF Millet Grain | | 0.8 | 1.3-3.4 | | (1.0) 0.6 |
| Stalks | | | 4.9-12.5 | | |
| RF Wheat | 1.2 | 1.5 | | | |
| Potatoes | 14.4 | 12.8 | | | |
| Tomatoes | 11.5 | 15.2 | | | |
| Qat RF | 974* | | | | |
| Qat SI | 1414* | | | | |

*Unit = Mandil/ha, RF= Rainfed, SI= Supplementary Irrigated.

Farm costs and incomes from the NWRA survey combined with the water use assessment from the preceding section and for the same area (Table 3.8) permits an analysis of water use and costs. A rudimentary assessment of the costs of wells, irrigation and farm budgets was made during the 1995 and 1996 rural surveys. The latter data were within +/- 50% of the farm budget obtained by NWRA and are well within the variation noted in NWRA (1999a). Because the NWRA sample was much larger than this survey's, the former data have been used as the basis for analysis.

Based on the labour inputs, Table 3.8 indicates the relatively small number of people needed to produce the crops. This calculation assumes continuous full employment for 270 days/year and is included to permit comparison with industrial returns to water discussed in the next section. In fact labour demand is seasonal. The minimum "man seasons" needed to produce the crops is 6200.

Table 3.8 also indicates the large proportion of costs incurred by irrigation relative to income. This is explored further in Table 3.9. Table

3.9 assesses costs and returns for the irrigation sector alone and also the affect of variations in irrigation costs for the agricultural sector as a whole:

1.  at current costs,
2.  if the price of diesel for the pumps were to increase to border parity levels, and
3.  if border parity price diesel and capital depreciation are included.

[The method of calculation of the current cost of water in Table 3.9 is different in "a) Running Costs" from parts b) and c). The former figure is derived from the data in Table 3.8 whilst the latter two are calculated using the assumptions given in Table 3.9. As a check of the validity of the latter method the "current water cost" was calculated for section a) using the method for b) and c) and resulted in a cost of 95MYR ($0.73M) which agrees satisfactorily with the former method (102MYR, $0.78M).]

*Summary of economic analysis* With only diesel charged at border parity levels, the profitability of irrigation is reduced almost to nil. [If farm costs or income were to vary by plus or minus 50%, as data error might tolerate, then irrigation would become unprofitable at current income levels.] Although current costs estimated by farmers do not include capital depreciation (and neither do World Bank estimates; 1993, Annex 2, Moench, 1997; 12), when these costs are included, the returns to water become negative. This conclusion is maintained even if field averages in Table 3.8 underestimate farm income or overestimate expenditure by 50%.

The argument that Yemeni farmers could grow irrigated crops at a loss becomes plausible only because much of the well digging/drilling and purchase of pumps and pipes occurred when remittances from Saudi Arabia and the Gulf were being invested in the farms prior to the Gulf War. Because of the omission of capital depreciation, the current running costs in Table 3.9 relate to the economic awareness of the irrigator which are much less than his "real costs". The sensitivity of the depreciation period was analysed by doubling the "lifespan" of the borehole / casing to 30 years and the pump and motor to 15 years. In this instance the irrigation returns to water indicated in the "real costs" part of Table 3.9 would be minus 5.2 $YR/m^3$ (-0.04$/m^3$) and average returns to water for the whole sector would be 1.9 $YR/m^3$ (0.015$/m^3$).

**Table 3.8 Agricultural expenditure, returns and water costs in the Upper Wadi Rasyan catchment**

| | Cultivated Area | | Total Costs[c] | Gross Income[c] | Net Farm Income | Net Area Income | Family Labour | Hired Labour | People Employed | Current[c] Water Costs | |
|---|---|---|---|---|---|---|---|---|---|---|---|
| | Wadi Irrigation | Non Wadi[b] | (including family labour but excluding water) | | | | | | | | |
| | ha | ha | $/ha | $/ha | $/ha | $/ha | days/ha | days/ha | 'man years'[e] | $/ha | $/yr |
| **Summer** | | | | | | | | | | | |
| Irrigated Qat[a] | 146 | 25 | 347 | 2595 | 2249 | 384512 | 22.6 | 40.7 | 40.1 | 310 | 53089 |
| Rainfed Qat[a] | | 1013 | 281 | 1788 | 1507 | 1526102 | 19.8 | 35.5 | 207.1 | | |
| Sorghum or Millet | | 7455 | 195 | 379 | 184 | 1539821 | 19.4 | 12.6 | 992.1 | | |
| Sup.Irrigated Sorghum or Millet | 258 | 72 | 249 | 389 | 139 | 45999 | 23.6 | 17.3 | 49.9 | 155 | 51186 |
| Maize | 146 | 966 | 299 | 504 | 205 | 228328 | 27.1 | 20.9 | 197.7 | | |
| Sup. Irrigated Maize | 648 | 108 | 283 | 489 | 206 | 155522 | 23.6 | 17.3 | 114.4 | 198 | 149732 |
| Fruit[a] (Irrigated) | 38 | | 313 | 1331 | 1018 | 38279 | 6.0 | 47.2 | 7.4 | 914 | 34358 |
| Vegetables (Irrigated) | 30 | 32 | 745 | 1030 | 285 | 17671 | 86.8 | 87.4 | 40.0 | 1229 | 76173 |
| Wheat / Barley | | 280 | 1109 | 692 | -418 | -116934 | 145.3 | 72.3 | 225.6 | | |
| **Winter (Irrigated)** | | | | | | | | | | | |
| Maize | 450 | 180 | 301 | 510 | 209 | 131757 | 28.9 | 17.2 | 107.6 | 278 | 175232 |
| | 128 | | 265 | 406 | 141 | 18060 | 28.9 | 17.2 | 21.9 | 218 | 27984 |
| Potatoes | 234 | | 1359 | 2658 | 1299 | 303892 | 27.6 | 61.7 | 77.4 | 296 | 69152 |
| Vegetables | 86 | 36 | 944 | 1269 | 325 | 39589 | 99.0 | 70.0 | 76.4 | 396 | 48342 |
| **Third Season (Irrigated)** | | | | | | | | | | | |
| Maize | 62 | 108 | 301 | 510 | 209 | 35554 | 28.9 | 17.2 | 29.0 | 278 | 47285 |
| Sorghum or Millet[d] | 156 | 72 | 265 | 406 | 141 | 32068 | 28.9 | 17.2 | 38.9 | 218 | 49691 |
| **TOTAL** | | | | | | 4,380,221 | | | 2225.5 | | 782,224 |
| Total Irrigated Agriculture Only | | | | | | 1,202,903 | | | 603.0 | | 782,224 |

[a] Perennial.

[b] 6% of non-wadi qat is assumed to be irrigated: 3% by tankers and 3% by boreholes. See text.

[c] Based on NWRA (1999a). For further analysis of water costs see Table 3.9.

[d] Pro-rata of summer cost and income relative to maize.

[e] Assumes 270 days worked per year and is 'man years' required to produce the crop.

Source: Fieldwork and NWRA (1999a).

## Table 3.9  Cost of water to agriculture (Upper Wadi Rasyan)

| a) Running Costs | Water Used | Current Water Cost | Current Water Cost | Net Income | Returns to Water |
|---|---|---|---|---|---|
| | Mm³/yr | M$ | $/m³ | M$ | $/m³ |
| Irrigation Only | 30 | 0.78 | 0.03 | 0.42 | 0.01 |
| Total Agriculture | 133 | 0.78 | 0.01 | 3.60 | 0.03 |

| b) Future Running Costs? | | Cost with diesel at border parity & no depreciation | | Net Income | Returns to Water |
|---|---|---|---|---|---|
| | Mm³/yr | M$ | $/m³ | M$ | $/m³ |
| Irrigation Only | 30 | 1.16 | 0.04 | 0.04 | 0.00 |
| Total Agriculture | 133 | 1.16 | 0.01 | 3.22 | 0.02 |

| c) Real Costs? | | Cost with diesel at border parity & with depreciation | | Net Income | Returns to Water |
|---|---|---|---|---|---|
| | Mm³/yr | M$ | $/m³ | M$ | $/m³ |
| Irrigation Only | 30 | 3.12 | 0.10 | -1.92 | -0.06 |
| Total Agriculture | 133 | 3.12 | 0.02 | 1.26 | 0.01 |

**Assumptions in calculating depreciation:**

Operating Period[b]     5.8 hrs/day, 6.5 days/week

| Borehole Items | Depreciation Period[e] | Price[d] | Average Depth[b] m | Number of Wells[b] |
|---|---|---|---|---|
| Casing & Riser -Dug | 15 years | 154 $/m | 15.3 | 1073 |
| Casing & Riser -Drilled | 15 years | 77 $/m | 150.0 | 237 |
| Pump and Motor | 10 years | 11500$ | | |

| | Current | Border Parity[c] |
|---|---|---|
| Diesel Price | 10YR/lit | 18YR/lit |
| Diesel Consumption[d] | 2.75 lit/hr | |
| Oil Expenditure[d] | 0.39 $/day | |

[a] Farmer Estimates of Water Costs (current diesel price & excluding depreciation).

[b] NWRA Well Inventory (1996).

[c] World Bank (1998; Annex 2 pg 3).

[d] Handley (1997; 6).

[e] See sensitivity discussion in text.

*Ten Year Shift in Irrigated Area*

Comparison of summer and winter satellite images of the Upper Wadi Rasyan catchment from the 1985/6 and 1994/5 suggests changes in agricultural water use. Due to cloud-free image availability it was not possible to get images from the same summer months for the two dates. One was for October and the other for June. For most of the summer cropped area this is not thought to be a problem since field work suggested there is only 1% difference in cropped area between June and October. One reason for the similarity is that a second maize crop has been planted and occupies nearly the same area but in different fields. Qat occupies the same locations and sorghum has not been harvested by early October. Apart from the differences in cultivated area noted in Al Hayma (chapter 2) between the mid-1980s and the mid-1990s the total area of cultivation in the major wadis increased by around 16% although the area under winter cultivation has declined by 38%. Areas of reduced cultivation are downstream from Al Hayma as far as Burayhi and Wadi Malih, from lower Wadi Dabbaab to Wadi Rasyan (see Figure 2.11 for locations). These declines coincide with the areas downstream of city and factory abstractions. Declines in winter cultivation are also noted in upper Wadi Maliha, Wadi Rubay'i and upper Wadi Sawaami. A decline in winter irrigation intensity in Wadi Dabbaab, and an increase in Wadi Hidran to Wadi Rasyan, lower Wadi Miliha and lower Wadi Sawaami are also evident.

*The Cost of Depleting the Al Hayma Aquifer*

Combining the assessment of crop areas from this study with the agricultural economics data of NWRA (1999a) for Wadi Al Hayma permits an estimation of the cost of returning the aquifer to its pre-development state (less the current abstraction for the city). The value of profitable irrigated agriculture and the labour lost if no irrigation were to take place until full recovery would amount to 1.8BYR ($14M) which includes the labour costs of around 1000 directly related lost livelihoods (Table 3.10). These returns to water are contrasted with industrial returns and are discussed further below.

## Table 3.10 The cost of environmental degradation
The loss of crops and jobs through not irrigating whilst the Al Hayma / Miqbaba aquifer refills

| | Area of Wadi Irrigation ha | Farm Costs[a] $/ha | Labour Costs $/ha | Gross Farm Income $/ha | Net Farm Income[b] $/ha | Net Area[c] Income $/yr |
|---|---|---|---|---|---|---|
| **Summer** | | | | | | |
| Irrigated Qat[a] | 109 | 445 | 226 | 2447 | 1776 | 193633 |
| SI Sorghum or Millet | 8 | 260 | 120 | 355 | -24 | -195 |
| SI Maize | 94 | 367 | 203 | 516 | -54 | -5067 |
| Coffee | 6 | 492 | 553 | 800 | -246 | -1475 |
| **Winter (Irrigated)** | | | | | | |
| Maize | 8 | 315 | 195 | 664 | 154 | 1233 |
| Sorghum   or Millet[d] | 5 | 222 | 115 | 457 | 119 | 595 |
| Potatoes | 190 | 1284 | 337 | 2695 | 1074 | 204099 |
| Vegetables | 10 | 511 | 623 | 1533 | 399 | 3987 |
| **Third Season (Irrigated)** | | | | | | |
| Maize | 4 | 315 | 195 | 510 | 1 | 3 |

**Total Irrigated Agriculture Only**

Annual Lost Income= 396,812
Annual Lost Profitable Activity= 403,550
Annual Lost Jobs= 1073

**Total Cost To Refill Al Hayma Aquifer (M$)**

| Basin | Years | Jobs | Agricultural Income | Total |
|---|---|---|---|---|
| Al Hayma | 9 | 8.42 | 3.05 | 11.47 |
| Miqbaba | 4 | 0.71 | 0.26 | 0.97 |
| Total | | 9.13 | 3.31 | **12.44** |

[a] Excluding Labour and Including Water.
[b] After Deducting Labour Costs.
[c] Al Hayma - Miqbaba Valley.
[d] Pro-rata of summer cost and income relative to maize.

Conversion rate $1 US = 130 YR.
*Sources*: Fieldwork and NWRA (1999a).

**Industrial Sector Water Use**

The development of industry in Yemen has been steady and the sector is approaching the share of GDP enjoyed by agriculture (Figure 3.2). In the Yemeni context the Upper Wadi Rasyan catchment of the Ta'iz governorate has been at the forefront of industrial development with the larger companies alone accounting for an estimated 14% of the national workforce compared with a 4.4% share of the population (Ta'iz Chamber of Commerce figures CSO; 1997; 39, 20). This assumes that the major factories of the Upper Wadi Rasyan area account for 80% of the industrial activity in the governorate.

*Methodology*

During the summer of 1995, visits were made to the larger factories in the immediate vicinity of Ta'iz. Several different sources concurred that these factories accounted for between 90 and 95% of the industrial water users that do not depend on the public supply. Some of the larger factories were revisited on occasions between 1996 and 1998 as opportunity permitted. Although coverage of industrial water users was high, the number of factories using that water is small and precludes statistical analysis of responses. Because of the size of the sample and the nature of the questions, key informant interviews were conducted with technical and managerial staff. Checklist questions were used covering the following:

1. Number of employees.
2. Attempt to assess the total amount of water used.
3. Sources of water and means of abstraction and distribution.
4. Costs involved in supplying water.
5. Uses of water.
6. Means of waste disposal.
7. Problems encountered.

The purpose of the questions was to obtain data permitting assessment of:

1. water uses and problems,
2. the financial and livelihood returns to water, and
3. awareness of and willingness to address equity and environmental issues in source and disposal areas.

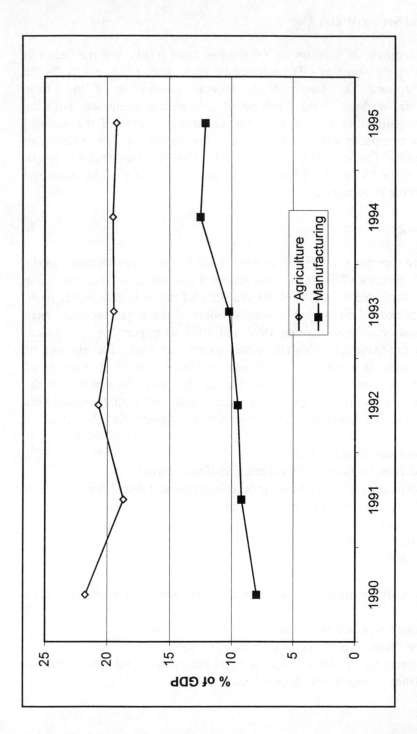

**Figure 3.2  Sectoral share of Yemen's GDP**   *Source*: Statistical Yearbook, 1995.

Technical staff were usually very willing to give a conducted tour of the source and disposal areas as well as the main water use points in the factories. The tours permitted some contact with the inhabitants of the source and disposal areas. If not contacted during the tour, then an attempt was made to meet with them on another occasion. Apart from the observations given below, it was noted that managerial staff were less aware of the problems of the water supply and disposal systems and environments than the technical staff and that the technical staff were less aware of the equity issues involved. Technical staff were also less aware of any steps taken to remedy or contribute to problems related to those issues.

In assessing returns to water, data has been obtained from national industrial statistics and applied on a pro rata basis according to workforce. It is recognised that company income may be falsified to national bodies for tax reasons, but they might also be to a foreign researcher, such as the author. To ensure the rest of the questions were answered willingly, it was not considered information effective to pry into financial aspects.

*Observations*

*Water use* Most of the major water consuming industries of Ta'iz are involved in the processing of foodstuffs and drinks, however the soap and ghee and paint factories also use substantial amounts. As well as using water in the products, the factories use water in the manufacturing processes, particularly for cooling, for steam production and in cleaning equipment and surfaces, and in general consumption by employees.

The main water related problems noted by the companies were firstly inadequate water quality for the industrial processes involved, particularly high levels of Total Dissolved Solids and hardness leading to scale problems in the cooling systems. Problems with particular constituents, such as alkalinity, sodium, calcium and magnesium and bacterial content were mentioned. Some factories faced problems of water scarcity, particularly where their own wells did not provide water of adequate quality and they have to bring water by tanker. Queues at the supply wells have resulted in delays and loss of production, especially during dry periods. Problems of waste water disposal and conflict with surrounding rural communities over abstractions were also noted and these aspects are discussed in later sections.

National statistics (CSO, 1997) indicate that the following proportions of larger industrial establishments (that is, of more than ten employees) face difficulties due to rising water costs (18%), continuous cut off of water (13%) and lack of water from the public network (26%). Due to the public water supply problems, it is considered that the proportion of the latter category in the city of Ta'iz would be even larger than the national average. Table 3.11 summarises the water use and employment data for the major water consuming industries of Ta'iz.

Proctor and Gamble, NCSPI and Bilqis Plastics and Red Sea Detergents account for around another 140m$^3$/day water consumption (Haskoning, 1990) and from a brief survey of stone cutters, a further 275 m$^3$/day of water is used in their work. These industries also depend on either their own wells or on tanker deliveries. CES (1997) estimate that around 20% of the public supply is used for small industrial and commercial companies. If the industrial component comprised half of this, that is, around 1800 m$^3$/day, the estimated total industrial consumption for the Upper Wadi Rasyan area would be 5230 m$^3$/day +/-15%. This total includes 150 m$^3$/day imported into the area from the Wadi Warazan catchment and 7 m$^3$/day exported from the area to Al Mocha. A total industrial water requirement of 150% to double the amount mentioned above has been calculated (As Sayagh; 1998) however the calculation is based on standard formulae and not fieldwork.

The annual value of production and profits of the industries in the Upper Wadi Rasyan catchment, assuming the national average and estimated pro rata rate on the proportion of the national workforce, are 20,900MYR and 7,950MYR respectively. Unlike the agricultural water users described in the previous section, industrialists include the cost of providing water within their accounting in the long-term, however the allocation of funds for future investment tends not to be addressed until a "water problem" arises. On a day to day operational basis, the only aspect of water costs that managerial and technical staffs were aware of how much their tankers had to pay at the well, or, if they received water from the public utility, how much the monthly bill was. Although more data were available to them in their accounts departments, there was no awareness of the actual costs of water in terms of costs of well, pump, pipes and tanker depreciation, staff and running costs. There was, however, a general awareness amongst most technical and managerial staff that water is a costly facet of production especially if production was lost due to the constraint of water supplies. Water conservation measures, including training of staff in water conservation and investigation of water reuse potential were being undertaken by the two larger groups of companies.

**Table 3.11 Industrial water use in Ta'iz**

| Company | Water Use | Employ- ees | Consump- tion m³/day | Jobs/m³/ day |
|---|---|---|---|---|
| Soft Drinks | Drinks | 32 | 175 | 0.2 |
| Bottled Water | Drinking Water | 37 | 200 | 0.2 |
| Soap & Ghee | Processes | 900 | 1000 | 0.9 |
| Nadfood (Food and Drink Products) | Product and Processes | 850 | 900 | 0.9 |
| Ice | Ice | 10 | 10 | 1.0 |
| Paradise Juice (Juices) | Drinks and Processes | 100 | 60 | 1.7 |
| Shebani Food (Food Products) | Product and Processes | 700 | 140 | 5.0 |
| Genpak (Food and Packaging) | Processes | 1100 | 220 | 5.0 |
| Paint | Product and Processes | 160 | 25 | 6.4 |
| YCIC (Food Products) | Product and Processes | 1850 | 285 | 6.5 |
| Total | | 5739 | 3015 | |

Assuming that the Upper Wadi Rasyan area accounts for 80% of the industrial activity in the governorate, the total direct employment in the industries would be 12,700. It is considered that this figure accurate to within +/-5%.

*Comparison with agriculture* A comparison of financial and livelihood provision returns to water between irrigated agriculture and industry provides a stark contrast (Table 3.12).

These figures are based on current operating costs for agriculture. When depreciation on boreholes and pumping equipment is taken into account the returns to irrigation water become negative and the contrast is even more stark. The errors built into the above values through the assumptions involved in calculating them do not detract from the conclusion that industry makes far, far better use of water than irrigated agriculture in both income generation and livelihood terms. The trickle-down benefits these industries impart to the service industries would make the contrast greater again.

**Table 3.12   Returns to water: irrigated agriculture vs. industry in the Upper Wadi Rasyan catchment**

| | Water Used Mm³/yr | Net Income MYR/yr. (M$/yr.) | Returns to Water YR/m³($/m³) | Jobs Provided | Jobs/ Mm³/yr |
|---|---|---|---|---|---|
| Irrigation | 30 | 29.08 (0.22) | 1.8 (0.014) | 600 | 20 |
| Industry | 2 | 7950 (61) | 4180 (32) | 12700 | 6350 |

*Industrial waste water disposal*   None of the factories visited measured effluent quantities, and management and technicians had no idea of the quantity of their waste water or its concentration and constituents, although the larger groups were aware that pollution is a problem. Surveys by Haskoning (1990) and Water Care Associates (1995), instigated by the Environmental Protection Council and the Hayel Sa'id group respectively, of half the factories included in this study's survey indicated that around 65% of the water supplied to the factories was returned as waste. This would suggest a total of 4400m³/day waste water of which at least 75% is not disposed of in the sewerage system.

Apart from an inadequate aeration treatment works at Hayel Sa'id's Nadfood site and a treatment works purported to have been completed at the same group's Soap and Ghee factory since this survey was competed, none of the factories treat their waste water. Disposal is typically to lagoons or wadis near to the factory although Nadfood and Soap and Ghee pump their "treated" waste to more remote locations. An assessment of the environmental impact of waste water disposal in the Upper Wadi Rasyan catchment was given in chapter 2.

*Equity issues*   Because the larger industrial plants are located near agricultural land, they affect the downstream rural communities. Depending on the exact locations the neighbouring rural communities are affected by both abstraction for the factories and by their waste water disposal. In some instances complaints about the smells near the discharge areas have led industrialists to pipe their waste further and further away from their factory, simply transferring the problem to another area. Some industrialists recompense farmers for loss of crops at the discharge point and offer to buy their (now polluted) land. It has been suggested that the polluting of land can be used as a tool to force people to sell it. The farmers expressed feelings of powerlessness in this situation.

Some farmers think the location of industrial discharge near their land is a convenient source of water for irrigation and appear oblivious to the dangers of degrading their soil and toxins entering the food chain. Where water is abstracted some factories try to compensate the affected rural communities by providing jobs, electricity and water, and defaulting on compensation has resulted in conflict in some instances. In other instances the provision of services by the factories has led to opportunistic behaviour by the rural communities. The equity issues raised by these situations are discussed further in chapter 4.

**Rural Domestic Water Supplies**

Access to water for domestic use in Yemen provides great contrasts, particularly between urban and rural water access, and forms part of the distinction between urban and rural environments. Table 3.13 summarises water supply and sanitation provision in Yemen and in the Ta'iz governorate.

**Table 3.13  Types of water supply and sanitation in Yemen and the Ta'iz governorate**

| | National Average % | National Rural % | National Urban % | Ta'iz Avge % | Ta'iz Rural % | Ta'iz Urban % |
|---|---|---|---|---|---|---|
| Water Supply | | | | | | |
| Public Piped Supplies | 21.1 | 6.4 | 67.5 | 17.0 | 3.8 | 79.5 |
| Co-operative Project | 11.9 | 11.9 | 11.6 | 12.1 | 13.2 | 6.6 |
| Private Project | 5.9 | 6.0 | 5.5 | 3.3 | 3.2 | 3.6 |
| Well | 37.7 | 45.8 | 12.7 | 48.6 | 57.2 | 7.9 |

**Table 3.13   Types of water supply and sanitation in Yemen and the Ta'iz governorate cont...**

|  | National Average % | National Rural % | National Urban % | Ta'iz Avge % | Ta'iz Rural % | Ta'iz Urban % |
|---|---|---|---|---|---|---|
|  |  |  | Sanitation Provision |  |  |  |
| Sewerage Network | 10.5 | 1.2 | 39.8 | 11.7 | 1.2 | 61.2 |
| Covered Pits | 22.7 | 15.9 | 44.1 | 23.3 | 23.6 | 21.8 |
| Uncovered Pits | 18.2 | 21.9 | 6.7 | 25.9 | 29.9 | 7.1 |
| Nothing | 49.0 | 61.0 | 9.3 | 39.0 | 45.2 | 9.9 |

*Source:* CSO,1996b.

Nationally, the proportion of rural dwellings is 76% of the total whilst that in the Ta'iz governorate is 83%. Based on the national survey, the urban dwellers in the Ta'iz governorate are better served by piped water supply and sanitation than is common nationally, whereas rural inhabitants in the governorate have less provision of these services than the national average. [In the urban situation almost all co-operative and private water supply projects pipe water to the houses whereas in the rural environment the supply may be to a communal tank in the village or may be piped to houses.]

The rural survey described above included checklist questions regarding domestic water supplies and sanitation. The villages included in the survey demonstrated a wide variety of water supply sources and problems.

The villages of Al 'adan, Al Jahaaza, Al 'anjud, Ar Rahaybah and Al Malika have piped supplies (Figure 3.1 indicates locations). The schemes supplying Al 'adan and Al Jahaaza are called the Eastern and Western systems respectively. Both are supplied from wells drilled into the artesian Tawilah Sandstone in neighbouring valleys. They both comprise storage tanks located above the villages they serve which distribute via pipe to individually metered houses.

Permission to drill for the Western scheme was delayed due to disputes over water rights. It delivers to 4000 households and was established by local businessmen working in Saudi Arabia. Most people respond to demand management and stop using water from the tap when their consumption for the month reaches $6m^3$, at which point a higher tariff is applied. After this point the locals resort to women and donkeys to bring their water needs. Both the Eastern and Western schemes also charge 50YR/month (0.38\$/month) maintenance charge and although the Eastern scheme does not apply an increasing block tariff average, consumption remains at around $6m^3$/month per connection. The main problem facing the Eastern scheme is a lack of spare parts for the pump. This has caused the community to want to hand the scheme back to SURDU who constructed it. This scheme is one of the 92 constructed by SURDU in the Ta'iz governorate by 1986 which at that time served an estimated 220,000 people (FAO/World Bank, 1986: 23).

At 'adan As Safaa, the community-managed domestic piped water supply scheme was administered by a government-appointed shayx who used the income to his own ends. A subsequent village-appointed operator also took the revenues. In Ar Rahayba GAREW drilled the well and the local council and local shayx paid half each of the distribution system. The well supplying a private piped scheme to the upper, larger part of Al Malika has dried up, although six houses are connected to another well in the village. Everyone else has to fetch water. The shayx at Ar Riwaas paid for the village to be connected to the government (NWSA) supply and others from much further afield take water from it when their wells dry up.

Where households are not connected to a piped scheme, women and children, with or without the help of donkeys, make two to three trips per day to the well. Because jerry cans are considered too expensive (150YR, 1.15\$ for 20lit) and can get punctured by thorns, people often collect water in ex-corn oil containers. NWRA (1999b) estimate that on average women walk 1.7km to collect water each day and spend 2 hours walking and waiting their turn at the well.

Assessments of rural water consumption for domestic purposes consistently suggest 20l/c/d even where there is a communal piped scheme (Mullick, 1987, Handley, 1996b and NWRA, 1999b). This would suggest a consumption of around $2.5Mm^3$/yr based on the 1995 Upper Wadi Rasyan rural population. In addition to drinking water, this study estimates a further 90 litres per household per day ($1.5Mm^3$/yr for the area) are collected for cooking and other household duties. This quantity is 70% of the NWRA estimate (ibid.). Over 80% of the NWRA survey respondents considered the amount of water available was inadequate and that they had to spend too much time collecting it (ibid.).

Although no householders interviewed in this survey boiled their water, when the EC exceeds around 1000-1500µS/cm water quality awareness causes them to change source for drinking purposes if feasible in terms of cost and transport. Most of the houses had some form of water storage facility and in some instances 90% of the village had galvanized steel tanks on the roof or adjacent to the house. Wealthier families and small businesses that depend on water (such as chicken farms) bring water by tanker. Water disposal was nearly always to a nearby open pit, minor wadi bed or covered pit.

## Urban Domestic Water Supplies

*A Summary History of Urban Water Supply in Ta'iz*

Water supply is a crucial factor, if not the crucial factor, in the development of any settlement. Records of water supply facilities in Ta'iz go back as far as its history. Neibhur in the seventeen-sixties (Hansen, 1983; 272) describes large excavations to facilitate rainwater collection. Tradition has it that an aqueduct brought water from Jabal Sabir to the ancient water tank on the west side of Al Qaahira. Water channels dug into the side of the mountain are still evident lower down. A qanat runs from the old Jewish area at the head of Wadi Al Madam into the footslopes of Jabal Sabir. Some think there is also an ancient qanat running from the foot of Al Qaahira due north under Tahriir through the centre of the city to the Imam's pool at 'usayfra.

The twentieth century has seen a huge population increase in the urban area in particular (Figure 3.3) and a commensurate increase in urban water supply problems. The search for solutions goes back at least to 1917 when Cornwallis et al reported that the population of 4000 was supplied by piped water from Jabal Sabir. To this day the Muthaffar mosque continues to receive water from the foot of Al Qaahira and also from its own well.

The first major water supply network, the "Kennedy Memorial" system was constructed during 1962-1965 and was supplied from shallow wells in the Hawbaan area (Figure 2.1). At that time the population was about 60,000. The Hawjala wellfield was connected in 1970. The next major scheme was the construction of a water supply and sewerage system between 1977 and 1982, utilising the Al Hayma wellfield. The depletion of this field sparked off an emergency drilling programme in 1987 and 1988 immediately to the north of Al Hayma, with extension of the conveyor. Further depletion in Al Hayma resulted in piped water delivery frequency declining to once every 40-days and considerable water stress amongst consumers and at the offices of the public utility, NWSA. The most acute period occurred during summer 1995 and is often termed "the crisis".

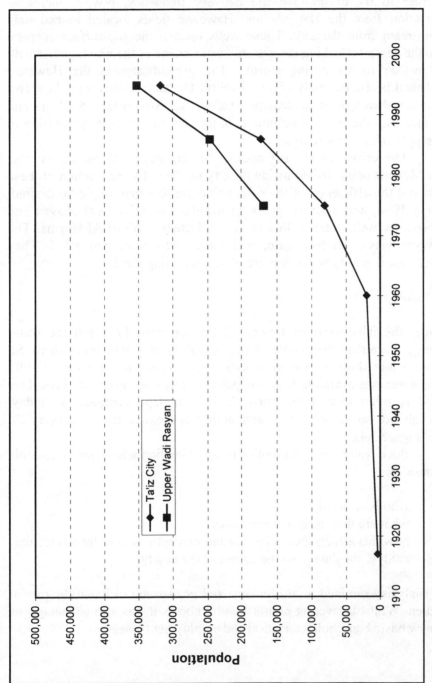

**Figure 3.3 Demographic trend: Ta'iz city and surrounding Upper Wadi Rasyan catchment**

In order to try to improve the delivery frequency, NWSA increased abstraction from the Hawjala and Hawbaan fields located immediately downstream from the city. These wells receive the sub-surface seepage from the cesspits, leaking sewers and water mains of the city together with the surface runoff during storms. The groundwater in the Hawbaan catchment is also naturally of poor quality. The abstractions from these two fields are therefore of inadequate quality, typically around 5000 $\mu$S/cm, and increases the level of pollutants in the city supply as its proportion in the supply mixture increases.

The crisis sparked off another "emergency" drilling programme, with 18 wells being drilled inside the city in 1995. The production of these wells and the efficiency of their operation are less than might be desired. During 1996, and after 10 years of negotiations with local shayxs and farmers, six wells were drilled in Habir, further north of Al Hayma. The conveyor was extended again, and three wells were connected. They provide good quality water, but are of disappointing yield.

*Methodology*

Besides the inhabitants of Lower Al Hayma, none have felt the water shortage more than the people of the city of Ta'iz – 400,000 of them. So far, the water shortage was most closely felt during the summer of 1995 when water was available from the public utility once every 40 days. This period is often termed the "crisis". The shortage continues to today, although not so acutely, with availability approximately once every 20 days (Figure 3.4).

Such water stress presents a unique situation and makes it possible to consider:

1. who is affected,
2. what are their adaptive responses,
3. how this affects their attitudes and perceptions to water-stress, and
4. analyse the data from the aspect of the equity.

The analysis examined wealth/income and gender and included the role of children. Whilst answering points 1 and 2 above, it was also advantageous to know basic household education and employment levels.

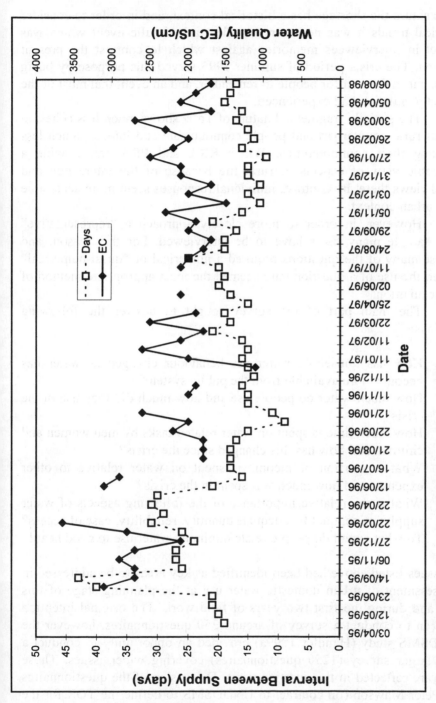

**Figure 3.4 NWSA water supply frequency and quality**

By its nature the shortage has a historical context, and in order to consider historical trends it was necessary to select a hydraulic event which was distinct in interviewees memories against which to contrast the present situation. The crisis period of summer 1995 served this purpose by being not too far in the past for people to remember, and an event that most of the present population had experienced.

The socially fragmented nature of Ta'iz shows much less cohesion than its rural counterpart and people communicate and interact much less regarding their commonalities. Where RRA and PRA can provide a reasonable view of aspects of rural life because of the interaction and shared views there, by contrast, individual responses seem more accessible in the urban context.

However, in order to more closely approach a "representative" urban sample many users have to be interviewed. For this reason, and because many of the questions required a numerical or "quasi-numerical" answer, the use of a questionnaire seemed the most appropriate method of data acquisition.

The main part of the survey sought to answer the following questions:

1.  How has household water use behaviour changed as water has become more available from the public system?
2.  How much water do people use and how much did they use in the crisis?
3.  How much time is spent on water related tasks by men women and children and how has this changed since the crisis?
4.  What proportion of income is spent on water relative to other expenses and how much was spent in the crisis?
5.  What is the relative importance of the following aspects of water supply: price, quality, adequate quantity, reliability, ease of access?
6.  To what extent do people relate minimum water use to good health.

The issues listed above had been identified as key areas to be addressed in an assessment of urban domestic water use at the planning stage of this study and during the first two years of field work. The original intention had been to conduct a survey of around 250 questionnaires, however the UNDDSMS study (Handley,1999a) provided an opportunity to conduct a much larger survey (1250 questionnaires) covering wider issues. Those issues are reflected in the questionnaires. Discussion of the questionnaires with Peter Mawson (on contract to UNDDSMS to define the TOR for the urban domestic water use survey) and Alex McPhail (World Bank) and input from them are gratefully acknowledged. A questionnaire was developed that would address the issues listed above, disguise income related questions, present the questions in a socially acceptable order and manner and be as brief, relevant and easy to comprehend as possible. Random sampling of 1028 houses was carried out with separate male/female interviews in 222 households. The sampling permitted analyses on the basis of income (Table 3.14) and tests for significant differences between male and female responses.

**Table 3.14 Skewed distribution of household income**

| Bracket | Income | Sample (No of Households) |
|---------|--------|---------------------------|
| 1 | Lowest | 149 |
| 2 | | 442 |
| 3 | | 215 |
| 4 | | 80 |
| 5 | | 52 |
| 6 | | 40 |
| 7 | Highest | 50 |

*Results Summary*

In reading this section it should be borne in mind that the analyses are no better than the quality of the answers to the questionnaire. Relatively poor correlation coefficients were obtained from the expenditure analysis (Handley,1999a), which was symptomatic of the quality of the numerical data. The tests for significance did, however, suggest that the following observations could be made.

**Table 3.15 Differences in female and male ranking of expenditure by gender analysis** (female rank minus male rank)

| Item | Rank | Item | Rank |
|------|------|------|------|
| Food | +0.52 | Water | +1.16 |
| Rent | +0.25 | Clothes | -0.11 |
| Education | -0.30 | Health | -0.44 |
| Qat | -0.05 | Transport | +0.33 |
| Electricity | +0.74 | | |

The proportion of expenditure on water as part of the items listed in Table 3.15 is around 4%. Per capita expenditure on water increases slightly with income and consumption may also increase slightly with income bracket.

The table suggests women think more is spent on water, electricity and food than the men think. Men think more is spent on health and education than the women think.

There are more adult readers and more employed people in the higher income groups, although the proportion of people in the household with jobs peaks in the middle income brackets. Men have received significantly more education than women at all education levels and incomes. 10% of the households sampled had no one with a job. Most of the households without employed persons were in the lower income groups. 10% of the lowest income group households had no adult readers.

The poor may be characterised in general terms as not owning land, houses, or a variety of consumer variables that are owned by the rich. Number of rooms in the dwelling also correlated strongly with income category. Most of the villas are occupied by those from the upper income groups. The axdaam (servant) caste forms a portion of the lowest income group. All those occupying temporary housing were in the lowest income group and many of them were axdaam. They typically have no waste water disposal facility. The lowest income group not only includes axdaam, however, and other poor families also were without any sanitation.

Other differences between rich and poor highlighted by the survey include the presence or absence of water-related facilities. Households not connected to the public utility tend to be poorer, and those connected to a private supply, wealthier.

*Water Use Patterns*

There are six main sources of water in the city:

1.  Public Utility, NWSA.
2.  Private Piped Supplies.
3.  Tanker Supplies.
4.  Free water (which can be obtained from government, private and mosque standpipes at various locations around the city).
5.  Bottled water (available from grocery stores).
6.  Jerry can water. This water is distributed to grocery stores where it is purchased as treated drinking water. It can also be purchased from the companies which treat the water and their branches,

however it is not referred to here as "treated water" because it may not have been treated.

**Table 3.16  Mean water prices from the different sources**

|  | Price YR/m$^3$ ($/m$^3$) | | % Using Source |
|---|---|---|---|
| Bottled | 26,000 | (200) | 4.0 |
| Jerry Can | 2,687 | (20.7) | 74.4 |
| Tanker | 285 | (2.2) | 33.5 |
| Private | 63 | (0.48) | 4.5 |
| NWSA | 24 | (0.18) | *84.0 |
| Free | 0 | | 35.2 |

* Best Estimate.

Water has to be obtained from all of the non-piped sources and this takes time. The time spent is related to income, poorest income bracket families spending nearly three times longer than those in the wealthiest bracket (Figure 3.6). A higher proportion of wealthy families obtain tankers, and the collection of free water is much more significant amongst poorer families (Figure 3.5). Poor families sometimes share NWSA connections in order to reduce connection fees. However this can result in water consumption being pushed up into the next block of the tariff and, in the end, both families pay more for their water. Water collection is mostly done by children. A higher proportion of children from poorer households collect water than from the wealthier households (Figure 3.7). Women think more time is spent collecting water and that more children and women are involved than the men think.

**Table 3.17  Percentage of men, women and children responsible for collecting water, according to men and women**

| Responsibility for Collecting Water | Men | Women | Children |
|---|---|---|---|
| According to Men | 48 | 3 | 40 |
| According to Women | 37 | 7 | 52 |

Although the number of women collecting water is much less than the number of men, the women's views are considered more realistic. All other water related tasks are almost entirely done by women, apart from watering gardens, which is evenly shared between men and women. Lower income households are more likely to keep livestock than wealthier ones and hence spend time watering them. In contrast, more wealthier families have gardens than poorer families, and have to spend time irrigating them. There is a significantly higher consumption of NWSA water for watering gardens amongst the rich. Although around 75% of all income brackets use jerry can water for drinking, the rich also use it for cooking.

18% of those using NWSA water for drinking boil it. Many fewer people drink free water or tanker water, and of those who do, many fewer boil it than boil NWSA water. Those without any piped connection resort more to free water for drinking and cooking purposes, and to a lesser extent jerry can and tanker water, than those with a piped connection. More households without a connection use free water and/or tanker water to a greater extent for water related tasks than do NWSA-connected households.

**Table 3.18    Number of households using different water sources for different household tasks**

| Source | drinking | cooking | bathing | washing clothes |
| --- | --- | --- | --- | --- |
| Jerry Can | 756 | 210 | 10 | 3 |
| Free | 113 | 261 | 306 | 307 |
| Tanker | 123 | 306 | 354 | 352 |
| NWSA | 203 | 718 | 905 | 915 |
| Private | 34 | 40 | 41 | 39 |

Since the water crisis of summer 1995, water use behaviour has changed significantly. Less time is spent now in collecting water, and people consider that more water is used in the home and that more time is spent using it. Despite the improvement in the public utility supply, total water supply from all sources still appears to be a little below the WHO recommended minimum (Figure 3.8).

Respondents were very discerning regarding water quality. Jerry can water was considered of good quality, and NWSA of poor quality (Figure 3.9). Other sources were placed between these extremes, but bottled and private water were considered better than free and tanker.

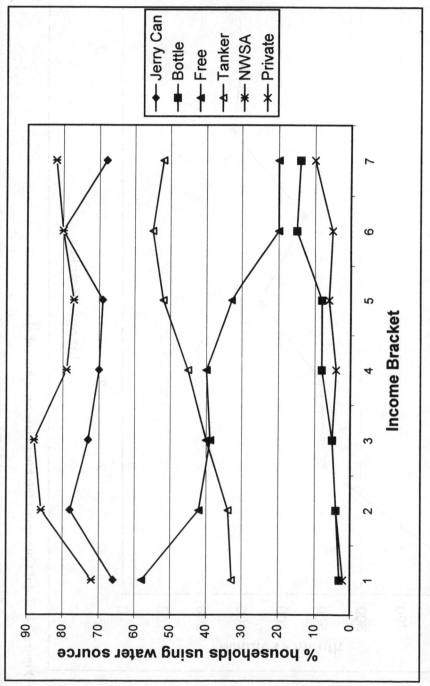

**Figure 3.5 Percentage of households in income bracket using water source**

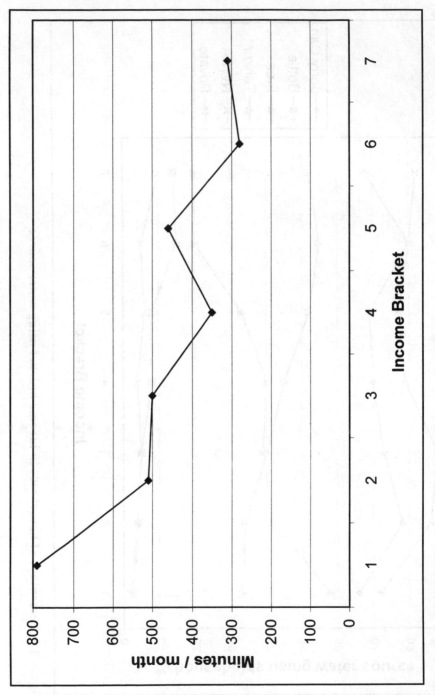

**Figure 3.6  Total time spent fetching water per household**

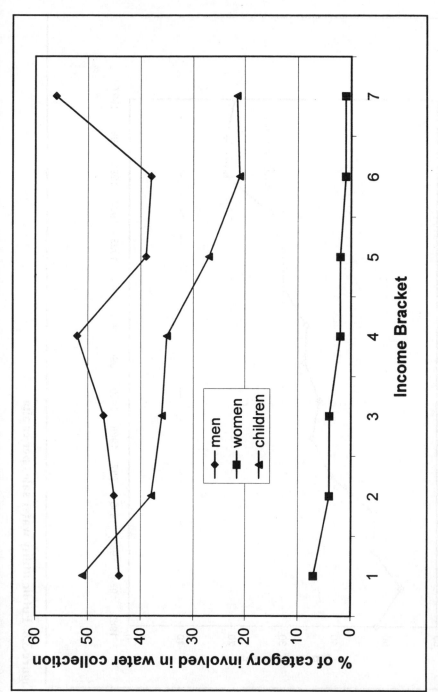

**Figure 3.7 Proportion of men, women and children involved in water collection**

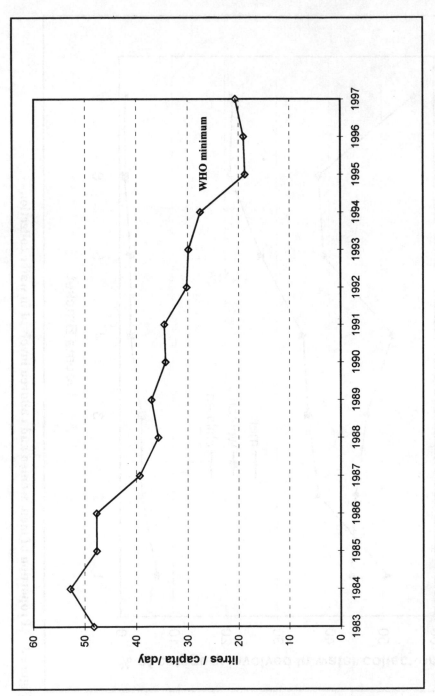

**Figure 3.8 Public utility water sales per capita**

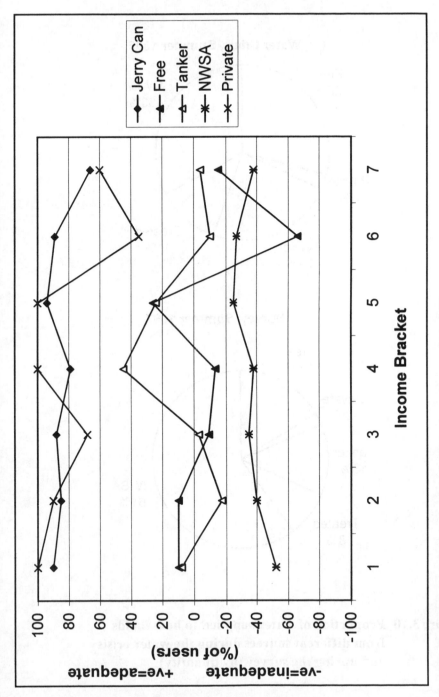

**Figure 3.9 Consumer views of adequacy of water supply quality**

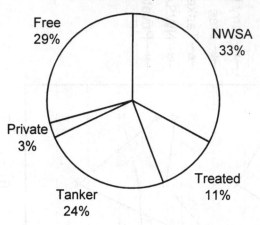

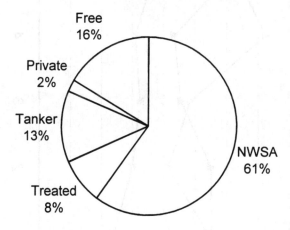

**Figure 3.10  Proportion of water supplied to households
from different sources during the water crisis
and during the survey (by quantity)**

Women had a higher preference for free water than men and men had a higher preference for jerry can and NWSA water than women. These preferences were based on both the quality and reliability of the water supply.

When specifically contrasting NWSA or private piped schemes, the main reasons for preferring either are reliability and price. However these reasons for preference become less important with increasing income, the wealthier preferring water sources on the basis of their quality.

**Table 3.19  Proportion of households that considered the quality of the water source adequate or inadequate**

| Response | % Using Source | % Adequate | % Inadequate |
|----------|---------------|------------|--------------|
| Jerry Can | 74.4 | 93 | 7 |
| Bottled | 4.0 | 95 | 5 |
| Free | 35.2 | 51 | 49 |
| Tanker | 33.5 | 50 | 50 |
| NWSA | 84.3 | 31 | 69 |
| Private | 4.5 | 89 | 11 |

Ease of access and water quality are more important for men (who, in urban areas, fetch the water) than for women, and the price was more important to women than to men. Women are less willing to pay for a regular daily supply of water than men are, and are also less interested than men in a private supply. More essential water uses, such as drinking and cooking, become less important to users in higher income groups and more important to lower income households and vice versa. The impact of current or future water shortages or increases in supply on health, cleanliness and convenience was considered less significant by the rich than the poor. 90% of the poorest category of those interviewed thought lack of water was detrimental to health. This proportion declined to 70% of the wealthiest category. The current water shortage was blamed by some for their health problems.

The water crisis of summer 1995 highlighted some of the problems and issues of water supply in consumers' awareness. Although the proportion of people utilising jerry can and tanker water supplies was the same as now, the amount of water obtained from them was significantly greater (Figure 3.10). More people were accessing free water, and were also accessing more of it, during the crisis. When the public utility supply failed to provide, many more people received a quantity of water in the flat rate part of the tariff structure. As consumption falls below $10m^3$/month (the maximum within the flat rate tariff), people end up spending more per $m^3$. Thus for $11m^3$/month the consumer paid around 270YR, 2.1\$ (water and sewerage) that is, 24.5YR(0.19\$)/$m^3$. If the supply rate falls below $5m^3$/month the cost per $m^3$ exceeds this. For instance $3m^3$/month at the flat rate of 120YR(0.92\$) minimum charge is equivalent to 40YR(0.31\$)/$m^3$. Thus, more is paid per $m^3$ when the water cannot be delivered!

Consumption of NWSA water was more strongly related to income during the crisis, with higher income groups consuming more than lower income groups, than is currently evident. Estimated consumption of NWSA water was reduced by around a third during the crisis relative to the time of the survey.

A lower bound estimate of consumption of NWSA water during the crisis should be based on NWSA records of water sold. The mean monthly NWSA water sales for 1995 was $196000m^3$ with an average delivery frequency of 21.5days giving $138350m^3$ delivery on each occasion. A 40 to 45 day delivery interval suggests approximately half this consumption. Assuming 20% is supplied for non-domestic use (CES et al, 1997), and 75% or 90% NWSA coverage, the household consumption from NWSA would have been 2.14 $m^3$/month or $1.78m^3$/month respectively, equivalent to 7.8 or 6.5 lit/capita/day.

*Conclusions*

There are large differences in wealth and opportunity between rich and poor in Ta'iz. Many of these differences are reflected in water use patterns and in attitudes and perceptions regarding water. Tanker, private piped, bottled and to some extent, public utility water sources are all accessed more by wealthier households and free water by poorer ones. However, the overall low level of supply, close to the WHO recommended minimum, has suppressed the demand for water. Although the wealthy have a much greater capacity for storing and using water, this capacity is not fully realised, and any future increase in supply is likely to be used disproportionately by this sector. Poorer households prefer water, from whatever source, primarily on the basis of price, and will spend much more time fetching it. Wealthy people are more interested in quality.

Most of the water provided by the public utility to the city supply operates on a system of recirculating polluted water. Water is pumped from the Hawjala wellfield to the city where, after use, it seeps from the cesspits and leaky sewers back to the well field. Low quality supply from the public utility has led to the sharp increase in the number of drinking water treatment companies. With 75% of the population drinking treated water, a dual supply system is effectively operating. Although people would prefer a high quality piped water supply, they are not willing to pay much more for it. In the absence of large amounts of high quality groundwater being discovered in the area, dependence on a dual supply is likely to continue for some time. People had to obtain more water from alternative sources during the 1995 crisis of the public utility supply. These sources (apart from free water) are more expensive. Because of the expense and the loss of convenience, people changed their water-related activities considerably.

Consumers are acutely aware of differences in time and money spent on the various water related activities and water sources as availability changes. They are also well aware of variations in water quality and to some extent of health implications of declining water availability and quality. The private sector has shown its ability to adapt and evolve to the need for improved quality drinking water and to provide in times of shortage.

## Water Markets

As indicated in the previous sections, water is widely bought and sold in the Ta'iz area, that is, water markets exist. It is the purpose of this section to examine their size and how they function. Three fundamental types of water market are identified:

**Table 3.20  Types of water market in Ta'iz**

| Market Type | Specific Markets |
| --- | --- |
| Rural - Rural | Irrigation (Described in agriculture section) |
| Rural - Urban | Public Piped, Tankers, Private Piped, Free Water |
| Urban - Urban | Jerry Can, Bottled |

Because of the distribution of water resources nearly all of the water used in the city is sourced from rural areas, however jerry can and bottled water could be considered separately since this is "produced" and sold in the urban setting.

*Methodologies and Descriptions*

*Private piped supplies* Each of the operators of the six private schemes operating in the city were interviewed. Each is essentially operated by one man and in total they supply not more than 230m³/day piped water to small, specific parts of Ta'iz. This estimate is considered accurate to +/- 20% and is based on tank sizes for the borehole schemes, and header reservoir sizes in the case of the spring-based schemes.

Four of the schemes are located on the slopes of Jabal Sabir where they tap spring sources. The declining yields of the spring-sourced schemes is considered to be due to the increased use of water upstream in the villages of Jabal Sabir for domestic purposes or qat irrigation. An increase in qat irrigation on the Northern slopes of Jabal Sabir during the period 1986-1995 has been confirmed by satellite image studies (Leung,1999). Although of good overall quality, the spring water is rich in fluoride, and accounts for the characteristic brown "Ta'iz teeth". Apart from in Saalah, households connected to the other schemes also receive NWSA water and are dependent on it for their non-potable supply. Yields from the schemes barely cover the potable water requirements.

Because of the limited supply, the spring water is used only for drinking, except for the Saalah scheme where there is no alternative non-potable NWSA supply. No new connections are being added to the spring-based schemes. The limited supply causes consumers even to supplement their drinking water needs with jerry can (dabba) water.

In all the schemes the customer arranges and pays for the physical connection between his house and the main, and no connection fee is charged. In each of the spring-based schemes, except Al Haadi, the customers are assessed by the operator for their ability to pay and are charged accordingly. A similar system operates when repairs are needed. In calculating the cost of production of the spring-based schemes, the reservoir and distribution systems have been valued at 1MYR ($7,700) and 1MYR ($7,700) per km of mains, discounted over 30 and 20 years respectively. All three schemes only allow for the operating costs including a small allowance for the operator and have no funds allocated for capital works replacement. The Saalah scheme is the largest, and operates as an essentially non-profit co-operative.

The Al Haadi scheme theoretically belongs to the 'awqaaf (religious endowment ministry) but no one pays for their water. Appropriation of water by force by some of the wealthiest sections of the Ta'iz community combined with a declining resource, has discouraged the person responsible for its upkeep and the system has fallen into disrepair, the number of connections declining from 520 to 120. All four spring-based schemes are running at a loss and, in the absence of maintenance, are steadily deteriorating. Of the two schemes based on well supplies in the city centre, one only serves the houses of the immediate family and the other sells water to about 50 houses at a profit.

In summary, the private piped schemes are in decline due to failing water resources and unsustainable institutional arrangements and form a negligible contribution to the water supply of Ta'iz.

*Private wells* Apart from the springs supplying the private piped supply schemes described in the preceding section, all water consumed in the city is derived from wells, most of which are located outside the city. Eighteen of the main wells that account for supplying at least 95% of the tankers were visited, and operators were interviewed regarding water sales. The main findings are summarised in Table 3.21.

**Table 3.21  Summary of survey of 18 private wells supplying Ta'iz**

| Well Data | Total |
|---|---|
| Discharge (lit/sec) | 75 |
| Water sold to tankers (m$^3$/day) | 1262 |
| Gross well income ($/day) | 255 |
| Profit less running costs ($/day) | 230 |
| Profit less depreciation over 15 years & less running costs ($/day) | 59 |
| Profit less depreciation over 30 years & less running costs ($/day) | -38 |

The well discharge was determined from the time required to fill a tanker. It is suspected that well operators may exaggerate the number of hours per day the wells are operating, and, for the wells outside the city, the operation time given in Table 3.21 includes the time the well is used for irrigation. For a variety of reasons the wells are not operated every day of every month. The well operator's estimate of running time is given without taking into account over-(or under-) estimates. Obtaining a reliable figure regarding the number of tankers supplied per day from well operator responses to survey questionnaires is also hazardous. Fear that the survey is for taxation purposes might lower the estimate, or a desire to be considered the owner of the most important and reliable well to prospective customers might raise the estimate. For these reasons the number of tankers per day collecting water from the well is only considered accurate to +/- 25%. Wells which do not have a storage tank of significant size in which to hold the water pumped from the well when tankers are not being filled are particularly limited in their delivery capacity. Wells are also limited in delivery capacity, even if they do have a storage tank, if they lack more than one delivery hose. The Ta'iz wells have neither storage tank nor more than one delivery hose. A well yielding the typical 5 lit/sec cannot turn round more than 50 tankers of $3m^3$ capacity in a 10 hour working day with no lunch or qat breaks.

There is no seasonal variation in price, although some wells increased the price slightly during the water shortage of summer 1995. The only wells yielding good quality water are in Wadi Ad Dabbaab and there is a slight recognition of this in the price paid by the customer. The income generated by sales to the tankers is indicated in Table 3.21.

In calculating the diesel and oil costs, it has been assumed that the motors use 200YR(1.5$) of diesel and 100YR(0.77$) of oil in a 12 hour day (well operator estimates, summer 1997 prices). This amount has been applied on a pro-rata basis according to the operating period calculated necessary to fill the tankers. Maintenance costs have been estimated at 20% of the depreciation. None of the private well operators do sufficient business to warrant employing others so no deduction for salaries has been made in calculating running costs. Although these assumptions are based on discussions with well owners, they should be considered rather approximate.

Table 3.21 indicates the difference between pre-depreciation profits versus the real loss all the wells are incurring when depreciation is taken into account. This difference is particularly marked for the drilled wells (those greater than 90m) where the construction costs are significantly greater than for dug wells. The dug wells are located outside the city and are also involved in agriculture (and, in some instances, block manufacture). To them, the income from the tankers represents immediate cash as opposed to the seasonal income from the farm. Many of the private wells were constructed with remittances from Saudi during the pre-Gulf War boom, a factor which could contribute to depreciation not being "felt" or accounted for by the operators.

Water quality is poor in most of the wells except those in Wadi Ad Dabbaab and there is no regular monitoring of the resource in terms of water levels and abstractions. The ability of the wells to maintain a service was severely tested during the water crisis, with some wells running dry. However they provide a vital supply line both for low quality water when the public utility fails and, more importantly, providing the main source of 75% of the population's good quality drinking water.

*Water tankers* Water may be obtained from vendors who usually own small tankers of approximately two to two and a half cubic metres. There are a few larger ones of four to five cubic metres. The tankers fill up at wells mostly outside the city and congregate at six centres. Each centre is organised into a rota by an elected, overseer who is paid by the driver. During August 1997, a survey of water tankers was carried out. Five tankers at each of the depots (a sample of 30% of the total number of tanker operators) were interviewed regarding:

1.  the wells they obtain water from,
2.  how much they pay for the water,
3.  where they deliver to,
4.  how many trips they make,
5.  how much they sell for,
6.  what their running costs are, and
7.  what their equipment costs new, how long it lasts, and how much maintenance it needs.

In most instances, each driver owns his own tanker and there are no indications of cartels in operation. The tankers deliver to the area surrounding their centres and are not usually asked to deliver further afield due to increasing cost with delivery distance. One exception is tankers which fill up with high quality water from Wadi Dabbaab, which may be asked to deliver to any part of the city. A summary of the survey is presented in Table 3.22.

### Table 3.22 Summary of survey of 30 water tankers supplying Ta'iz

| Tanker Data | Total (all centres) |
| --- | --- |
| Number of tankers operating | 103 |
| Number of tanker trips per day | 545 |
| Amount of water delivered (m³/day) | 1218 |
| Expenditure on tanker water by customers ($/day) | 2,440 |
| Tanker income excluding maintenance and capital costs ($/month.) | 66,000 |
| Tanker income including maintenance and capital costs ($/month) | 55,000 |

Table 3.22 gives averages for the five tanker operators interviewed at each depot. In calculating depreciation drivers estimate the new price of the

basic truck, tank, pump and hose at 1.1MYR(8500$), 50,000YR(385$), 60,000YR(462$) and 15,000YR(115$) respectively. These prices have been discounted over ten, five, five and three years respectively.

Running costs were obtained from the drivers for petrol and oil and averaged 4500YR(35$)/week and 2400YR(18.5$)/month respectively. No labour costs are incurred. Maintenance has been estimated at 20% of the purchase price spent over the depreciation period. After deduction of running costs, maintenance and depreciation, a very adequate income is provided if the demand for tanker water remains high (at five, or more, trips per day). When demand is high, tanker driver income is considerably more than the average income (as determined from the expenditure survey). However, when the NWSA supply improves, there is a prompt decline in work for tankers. To offset this risk, the tanker drivers can remove the tank and pump and become general hauliers instead.

The tankers are providing between 30% (for the lowest income group) and 55% (for the highest income groups) of the city with water. A major issue with the tankers is the water quality. Apart from incorrectly informing the customer regarding where they get the water from, tankers have been observed emptying cess pits and dumping it in the nearest wadi. Whether they are cleaned out before delivering the next load of "clean" water is not known. Although they are supposed to be cleaned out regularly drivers often bribe the municipality "baladiyya" inspectors not to bother them. There is, therefore, a need for legislation, spot checks, and for enforcement of quality standards.

Despite some shortcomings in the tanker service, they form a very effective safety valve to cover times of shortage. They rapidly adapt to changes in demand and maintain a round-the-clock service. They are also much more efficient at delivering water than the public utility, with little to no losses, and are more consumer aware.

They do not, however, cater for the poor, and, in fact, they typically prove more expensive for poorer clients. One reason for this is because there are proportionately more poor living in multi-storey apartments, and if there are more floors to lift the water up, tanker drivers charge more. A second reason is that many of the poor are located at the edges of the city. The outskirts are the main areas receiving new arrivals from the countryside and where land and rents are cheaper. These are the same areas which are not covered by the public utility and therefore need the tanker supply, but they are also further from the tanker depots and are often not served by asphalted roads, so, again, tanker drivers charge more to deliver to these areas.

*Drinking water treatment companies* All eighteen drinking water treatment companies were visited during the summer of 1997 and their operators interviewed regarding water supplies, processing, distribution and sales. Their responses to the interviews are presented in Table 3.23.

**Table 3.23 Summary of survey of 18 drinking water treatment companies supplying Ta'iz**

| Company Data | Total | Average |
|---|---|---|
| Operational Since | | 1994 |
| Tanker deliveries per day | 29 | |
| Water input (lit/day) | 180,400 | |
| Water sold by jerry can (lit/day) | 14,550 | |
| Number of branches | 56 | 3.1 |
| Water delivered by tanker to branches (lit/day) | 78,700 | |
| Total water sold (lit/day) | 93,250 | |
| Water losses | | 48% |
| Total sales value ($/day) | 773 | |
| Gross value of water sold ($/m$^3$) | 8 | |
| Profit after deducting running costs ($/day) | 412 | |
| Profit after deducting capital & running costs ($/day) | 49 | 2.7 |

Water is collected from good quality wells, either in Wadi Dabbaab or from Dhi Sufaal (near Al Qaa'ida) in owned or rented, dedicated tankers. The tanker brings water to the plant where it is processed first through a sand filter, then a fine cotton filter to reduce particulate matter and lastly through one or more Du-pont reverse osmosis (RO) cylinders to reduce salinity. For micro-organisms chlorine is claimed to be used. Although a chlorine bottle was often noticed at the plants, they were not generally connected because any hint of chlorine in the taste reduces sales. Some plants also add chemicals to the water in an attempt to precipitate salts and control pH. The key to processing the water is the RO equipment which is brought (or smuggled) from Saudi. The RO plant has some wastage, and ideally should be cleaned at periodic intervals to maintain efficiency. However this is seldom carried out, and rather than the manufacturers design wastage of 15%, the actual is around 50%.

A wide range of product quality can be found. Some operators manage to reduce the salinity whilst others actually manage to increase it.

In fact the sources are of reasonably good quality and were no treatment, except filtration, carried out, would produce an acceptable product for the public palate.

Water distribution is either direct from the plant via 5, 10 or 20 lit plastic jerry cans carried by pick-ups to the shops. Alternatively, many of the processors have branches and distribute to them by tanker (either the same tanker that brought the water, for low production plants, or another, usually smaller tanker, if the water-fetching tanker is busy).

The jerry cans may be owned by the drinking water treatment plant, the shops or the customer. Processing plants also sell a small amount of water direct to the public who live in the immediate vicinity. The branches contain a storage tank and a few taps for the public to fill from in a small room fronting onto a road. There are also various unofficial branches which receive water from one or more drinking water treatment companies. The water is sold for 15YR(0.115$) per 10 lit jerry can and 7.5YR(0.058$) per 5 lit to the shops and branches who in turn sell it for 20YR(0.15$) and 10YR(0.077$) respectively to the public. Jerry can water prices are relatively fixed throughout the city, though more is charged per litre for smaller quantities and if the water is cooled. The price has been monitored from summer 1995 to spring 1998. After steadily increasing, the price levelled out in summer 1996 (at 25YR(0.19$)/10 lit) falling for the first time in autumn 1996 with the seasonal decline in demand and it has not risen since. This may suggest that the market is now saturated as a result of the very rapid increase in the number of suppliers since 1994 (Figure 3.13). Some drinking water treatment companies have recently started to sell bottled water.

Consumption of jerry can water indicated by the household survey is eight times greater than the stated production of the drinking water treatment companies. There could be an "underestimate" of production by the companies for tax or other reasons. Drinking water may be also sold by vendors (shops or company branches) which is, in fact, untreated. This latter alternative is thought to take place rather often. The discrepancy may also result from households overestimating the quantities consumed, though this is thought less likely. Some water companies may have been overlooked in the survey, although efforts were made to include all the processing plants. The data given in Table 3.23 are based on stated sales and do not include the higher consumption of jerry can water indicated by the household survey. The amount of water stated as sold to the drinking water treatment companies by the wells included in the well survey should be less than that claimed to be bought by the companies since some water is also brought from the Dhi Sufal area which was not included in the well survey. The fact that the amounts are approximately the same suggests that the water processing companies have underestimated their water sales and purchases.

Companies were not readily forthcoming with information regarding equipment costs. In calculating the costs of the drinking water treatment companies, plant has been depreciated over ten years and vehicles over 15. Operating costs were estimated on the basis of a monthly rate of 20,000YR (154$) for building rental, 10,000YR (77$) for RO tube maintenance, petrol and oil at 22,000YR (169$) per vehicle, electricity at 3,000YR (23$) per month and salaries at 8,000YR (62$) per tanker per delivery per day. Other equipment maintenance was based on 20% of the depreciation rate. Only one rented building has been assumed for each processing plant. Often the branches are rented under a separate agreement by the branch manager who effectively operates as a financially separate business. If the water sales figures are to be believed, most of the drinking water treatment companies would operate at a slight loss if depreciation, calculated by the method used above, is taken into consideration. However, this loss is offset in some instances where the main building is owned rather than rented, by owning only one vehicle to collect and deliver water and by not maintaining the RO tubes. The latter is a false economy, resulting in the loss of up to 35% of their potential income. Company profitability appears to be rather variable, and depends on careful management and good technical know-how. Four companies have ceased to operate in the past two years probably through being unprofitable because of inefficient operation.

The market for jerry can water in Ta'iz is huge, with 75% of the city buying jerry cans, mostly daily. The water processing companies exist because the water from the public utility is perceived as inadequate for drinking.

Although there are generally few complaints regarding jerry can water quality, the fact that water sales stated by the companies are only one eighth of the quantity claimed to be purchased by the public suggests that much of the water bought has never been near a processing plant. Branches can be tempted to fill people's jerry can from the public supply, especially when they have not received a delivery from the processing plant.

Because of the potential for incompetence and cheating the public, monitoring and some command and control measures are needed. There is also a need for technical help in optimising the equipment operation. As with the tankers, there is no provision for the poor, and jerry can water is the largest part of the monthly water bill. Although the price has remained steady, because of the dependence of so many on the source, and, in the absence of significant finds of good quality water in the area, the consumer needs some kind of protection from profiteering. On the whole, the need for a potable water supply in Ta'iz has been met reliably and relatively cheaply by the water processing companies. Although the price of the drinking water is very cheap by global standards, it is still unaffordable to 25% of the population.

*The public water supply utility – NWSA* The supply of water to the urban population by the public utility NWSA differs from the other providers discussed above. Where this supply is different and where it is similar are worth noting. Regarding similarities, water is sourced from rural areas and the public utility competes with the other suppliers in a market for water in terms of price, water quality and reliability / ease of access. This competition is perhaps not immediately apparent in what is usually considered a monopoly. The monopoly, however, applies only to the piped nature of the NWSA supply (the private piped suppliers are almost insignificant in terms of their contribution to providing for the water needs of the city). Other differences between non-piped suppliers and NWSA relate to their ownership (private as opposed to public), and, when the water shortage is not too acute, the quantity of water they bring (that is, less than NWSA).

The estimation of per capita water consumption from NWSA supplies is difficult to assess for two main reasons. Firstly, the mean number of people per family in the census of Dec 1994 was 6.5, whilst the number per NWSA connection in this survey was 9.7. Secondly, the volume of water supplied from the water authority was $7123m^3$/day as estimated by NWSA versus $16878m^3$/day as estimated from household monthly water bills determined by the domestic water use survey. NWSA estimated 75% coverage in 1995 and 61% in 1998. These estimates are based on a simple model of one family ('usra) per connection and 6.5 persons per 'usra (1994 census figures and definitions). This assumes that a family is equivalent to a connection, that all houses are occupied and all connections legal. The random survey of 1028 households estimated 90% coverage and included illegal connections (2.5% actually admitted to this) and shared connections (another 4% admitted to this). The survey found an average of 9.7 persons per meter and excluded empty dwellings. An issue of whether the growth of the population of the city has been matched by a proportionate increase in the number of houses or whether there really more people per household is also raised by the conflicting persons per household statistics. All these factors lead to an underestimation of coverage by the NWSA method. Public utility underestimation of supply has been suspected from other surveys, such as San'a (Dar Al Handasah, 1997).

Similarly, the difference between the NWSA estimates of sewerage coverage (48%) and those of the urban domestic water use survey (68%) may result from a combination of illegal connections and also respondents assuming they were connected to NWSA sewers when in fact they are not. As well as cess pit sewerage, it is possible that there are some small private sewer networks, that is, a few houses connected to a common cess pit, or to the wadi, where there is no NWSA coverage.

**Table 3.24  Effect of assumptions on per capita consumption** (1997 figures of 382000 population and 35808 connections)

| | Scenarios | | | |
|---|---|---|---|---|
| People / Connection | 9.7 | 6.5 | 9.7 | 6.5 |
| % of total pop with connection | 91% | 61% | 91% | 61% |
| Water supplied (m$^3$/day) | 5700[a] | 5700[a] | 16878 | 16878 |
| Lit/capita/day | 16.4 | 24.5 | 48.6 | 72.5 |
| m$^3$/capita/yr[b] | 5 | 8 | 16 | 35 |

[a] NWSA figure and assuming 20% sold to institutions & commercial enterprises (CES,1997).
[b] WHO minimum 10m$^3$/capita/yr.

Regarding the quantity of water supplied by NWSA, the figure of 16878m$^3$/day is based on household bills. However there is a possibility that NWSA is over-charging, as indicated by the average charge of 24YR/m$^3$, equivalent to an impossible 15m$^3$/month delivery. Together with an average storage capacity of around 3m$^3$ indicated by the survey, the argument leans towards the lower quantity of water delivered and is likely to have been below the WHO guideline for some time (Figure 3.8).

As well as the problem of insufficient water supplies from NWSA, there is a problem of water quality. Although both aspects deteriorated during the 1995 water crisis and have improved somewhat since (Figure 3.4) the quality in terms of Electrical Conductivity remains worse than 1500 µS/cm.

*Free water* There are two main sources of free water in the city of Ta'iz. One is to collect rainwater. This is income related with a higher proportion of poorer households collecting rainwater (Handley,1999a). A second source of free water is obtained by queuing for water at the mosques, public standpipes and outside the houses of a few wealthy households that make water accessible to the public. The time involved in collecting free water (Figure 3.6) suggests that the poor spend time to save money and the rich vice versa. The trade-off between time for money indicates that "free" water is actually a market and the water carries an opportunity cost at least. For this reason it is suggested that free water should not become easy or quick to obtain, otherwise those in less need of it will use it. Free water quality can be very variable and, because it is free, be the least amenable to quality standards. Its users, the poor, are hence the most vulnerable.

*Water Market Overview*

Besides the estimates of consumption of NWSA-supplied water given in Table 3.24, non-NWSA water consumption is estimated in Table 3.25.

**Table 3.25   Quantities of water consumed by the city population from non-NWSA sources**

|  | Supplier Estimates $m^3$/day | Consumer Estimates $m^3$/day |
|---|---|---|
| Jerry can purchase by households | 93 | 790 |
| Water tankered to water processing companies | 180 | |
| Water sales to tankers by wells | 1262 | |
| Tanker sales to houses | 1218 | |
| Water purchase from tankers | | 2318 |
| Free water consumption | (1548) | 1548 |
| Total | 2859 | 4656 |
| Per capita, based on population of 382,000 at the time of the survey. | 7.5 lit/cap/day 2.5 $m^3$/cap/yr | 12.2 lit/cap/day 4.5 $m^3$/cap/yr |

The per capita water consumption lies somewhere between the supplier and consumer estimates and, because it is not possible for NWSA to provide the amounts indicated by consumer billing, the actual consumption is considered to be nearer the supplier estimate (16-24 lit/capita/day). This quantity is very low and when added to the provision from other sources only approaches the WHO recommended minimum (28 l/c/d). This quantity is not considered unlikely and similar estimates have been deduced for other Yemeni towns (for instance Amran and Yarim, Handley,1997a and Davies 1997 pers com.). Despite the discrepancies in the estimates, the importance of non-NWSA water sources are apparent (Figure 3.10). Alternative water supplies to NWSA comprise up to 37% by volume of all domestic water use

NWSA failure to provide water of better quality in salinity terms has led to most of the city depending on a separate supply for drinking water. This need has been met by a rapidly growing jerry can drinking water market. Although the volumetric quantity of jerry can, bottled and private piped water is not great, it does comprise over 85% of the drinking water. Around 20% of the non-potable water used is tankered water, which is the direct alternative to NWSA water. The proportion of tankered water increases above this level when NWSA production declines. Those who do not have the option to access alternative sources tend to be the urban poor.

Based on consumer estimates of household expenditure the average cost of water is calculated at just over 4% of household income. The poor pay a much higher proportion than the wealthy (Figure 3.11) and 89% of the population pay more of their income on water than the World Bank recommended maximum of 2% (quoted in Allan, 2000; chapter 3). Table 3.26 and Figure 3.15 summarise mean costs of water production from the different suppliers, inclusive of depreciation, and sale prices.

Based on summer 1998 data, and the national tariff applicable at that time, the price elasticity of demand for NWSA water in Ta'iz is −6 (Figure 3.12). The national tariff was 25.5YR(0.20\$)/m³ water and sewerage charge for 0-10m³/month water consumption, 42.5YR(0.33\$)/m³ for 11-20m³/month, 68YR(0.52\$)/m³ for 21-30m³/month and 85YR(0.65\$)/m³ for more than 30m³/month. The average monthly charge was 24YR(0.18\$)/m³ (Table 3.26).

Although private piped supplies and NWSA simply run at a loss, the drinking water treatment companies have mostly been established since the Gulf War brought a virtual end to the remittances (Figure 3.13). The other private water suppliers make an immediate profit that they live off and chew off, but it would need another remittances boom for significant replacement of plant and equipment or reinvestment.

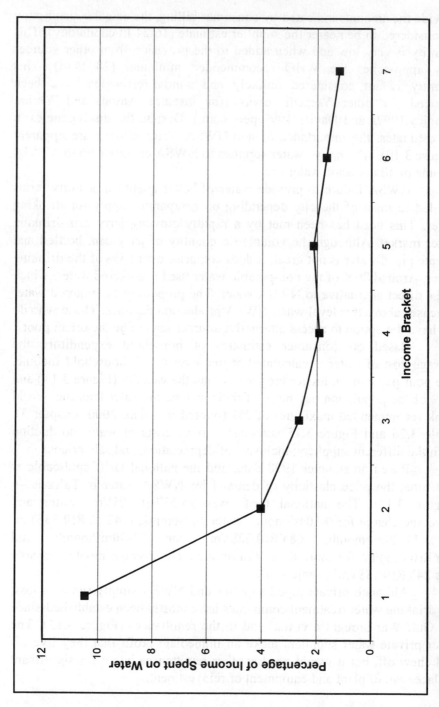

**Figure 3.11 Household expenditure on water as a percentage of income**

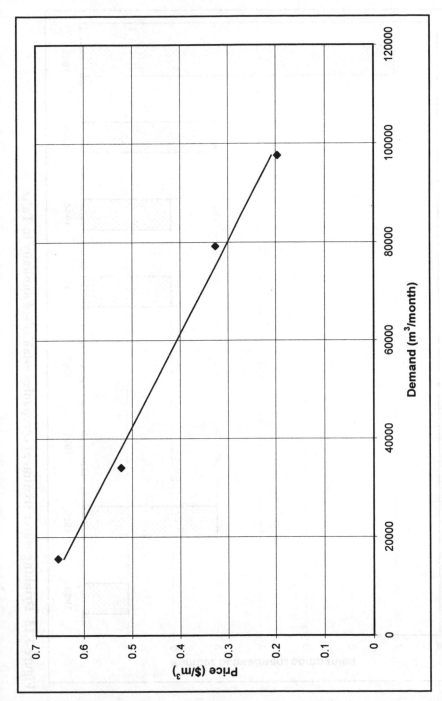

**Figure 3.12 Price elasticity of demand for NWSA water**

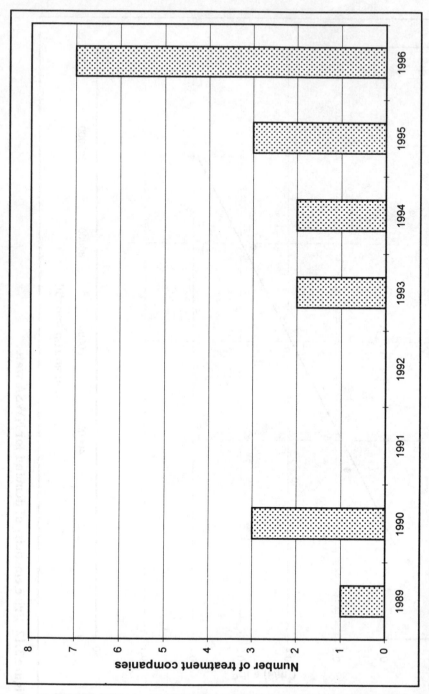

**Figure 3.13 Drinking water treatment companies established annually in Ta'iz**

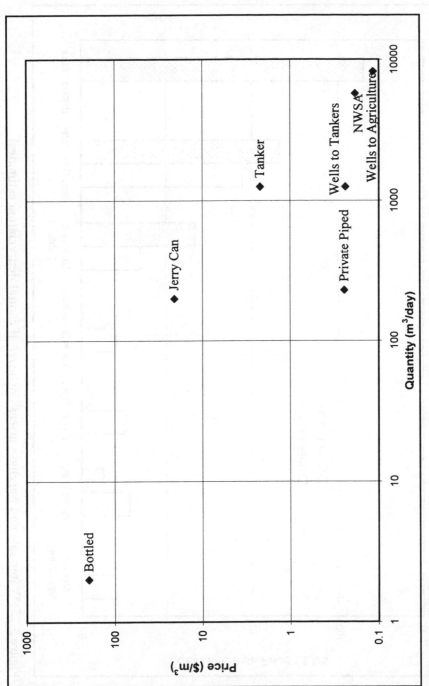

**Figure 3.14 Sale price and estimated quantities of water sold in Upper Wadi Rasyan water markets**

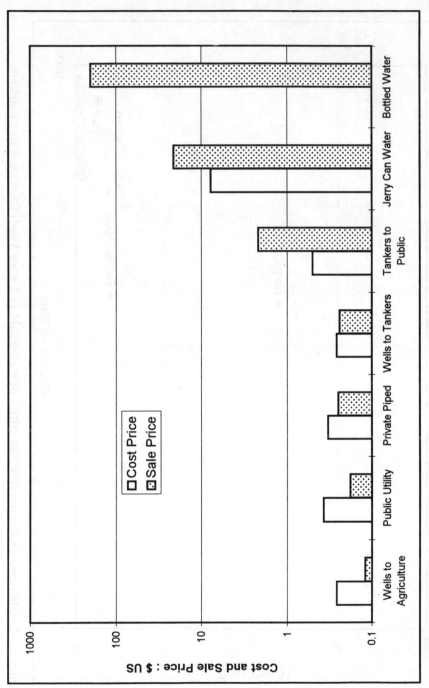

**Figure 3.15 Water sale and purchase prices (inclusive of capital depreciation estimates)**

**Table 3.26 Cost and sale price of water for various water sources**

|  | Cost Price YR/m³ ($/m³) | | Sale Price YR/m³ ($/m³) | |
|---|---|---|---|---|
| Wells to Agriculture | 33 to 35[a] | (0.25-0.27) | 15 | (0.12) |
| NWSA Supply | 48 | (0.37) | 24 | (0.18) |
| Private Piped | 43 | (0.33) | 32 | (0.25) |
| Wells to Tankers | 33 to 35[a] | (0.25-0.27) | 31 | (0.24) |
| Tankers to Public | 65 | (0.5) | 285 | (2.2) |
| Jerry Can Water[b] | 1,017 | (7.8) | 2,687 | (21) |
| Bottled Water | ? | | 26,000 | (200) |

[a] Higher cost for shorter depreciation period of well and equipment.
[b] Consumer estimates of purchases, not drinking water treatment company estimates of sales.

The production, sale price and water quality from each source are contrasted in Table 3.27 and production and price in Figure 3.14.

**Table 3.27 Quantity, quality and sale price of water**

|  | Quantity m³/day | Quality µS/cm | Price YR/m³ | Price $/m³ |
|---|---|---|---|---|
| Wells to Agriculture | 8200 | 1000 | 15 | 0.12 |
| NWSA | 5700 | 1750 | 24 | 0.18 |
| Wells to Tankers | 1250 | 3500 | 31 | 0.24 |
| Private Piped | 230 | 800 | 32 | 0.25 |
| Tankers | 1250 | 3500 | 285 | 2.20 |
| Jerry Can Water | 200 | 600 | 2687 | 21.00 |
| Bottled Water | 2 | 400 | 26000 | 200.00 |

The relationship between quantity supplied and water price for most sources lie on an approximately linear trend (on log-log scale). Notable exceptions are the private piped supplies, which now provide very small amounts of water and the sales by wells to tankers that do not cover their costs. Table 3.27 suggests the quality of water sold by the tankers to the public does not warrant the price charged. The loss of economies of scale by this method does not permit significant price reductions for tankered water and the lack of availability of wells providing good quality water from all but Wadi Ad Dabaab sources precludes an improvement.

## Summary and Observations

*Agricultural Water Use*

Rainfed agriculture topped up by spates in wadi areas and small areas of stream-fed agriculture was the traditional form of cultivation until the mid 1970s. Since then, the technological impact of boreholes and pumps financed by remittances from Saudi Arabia and the Gulf has led to the depletion of groundwater resources and hence also the depletion of spates and stream flows.

Despite the growth in groundwater development, the same period has seen a decline in the agricultural proportion of GDP and a population increase even in both urban and rural areas. The result has been a sharp decline in per capita water availability.

Land holdings are small and fragmented which forms a major cause of disputes. Relative to their urban counterparts the rural population is poorer. Rural wealth differentials are accentuated by water allocation which determines both the number of cropping seasons and the type of crops that can be grown. The wealthiest are those who control the headwaters of streams and groundwater flows and water resource development moves progressively back to this point. In the few locations where surface flows still occur, many have been polluted by urban and industrial development.

Areas of volcanic rocks are not very productive and alluvial tracts dominate the agricultural economy. The conflict over declining groundwater levels is concentrated in wadi areas at the base of the alluvium / top of the volcanics. As the groundwater level declines below this point only the few farmers with deep enough drilled wells can obtain water and maintain two or three cash crops per year at this inequitable final stage in the tragedy of the groundwater commons (Table 3.3). The direct cost of ultimate groundwater depletion has been evaluated for Al Hayma.

Since there is more cultivatable land available than water resources, improvements in production efficiency are only likely to result in more use of land rather than conservation of water. Although the irrigation methods are inefficient the main loss is in (cheap) diesel as water is recirculated to the aquifer. Efforts to reduce irrigation inefficiency are more related to the potential that the topography gives for the installation of piped distribution systems than to any other factor.

Although two restricted areas have seen a decline in agricultural production over the past ten years due to abstractions of groundwater for industrial use, declines have also been noted in non-industrial areas. A decline in rainfall over the period is possible but statistically unprovable.

Rainfed agriculture accounts for around 75% of the agricultural water use in the study area. Because of its distribution, this water cannot be used for any other purpose. The water derived from stream flows accounts for less than 10% of the water use and is difficult to appropriate for any other purpose since much is polluted and the remainder falls within strict allocative rules. The only water which could be reallocated to other sectors is the groundwater which comprises around 30Mm$^3$/yr in the Upper Wadi Rasyan catchment or 22% of the total agricultural water use. Even this modest amount includes a portion of overexploitation. Hence, right at the outset of this discussion, a severe limit is set on the potential for water reallocation.

Considering that economic and livelihood returns to water in agriculture are meagre and, if depreciation on boreholes and pumping equipment is considered, returns are definitely negative, one might ask if remittances were really a good thing? Despite these realities the Yemeni's heart remains in his soil.

## Industrial Water Use

Although Yemen is a relatively unindustrialised country, the manufacturing contribution to the GDP is increasing and, compared with the rest of the country, the Upper Wadi Rasyan catchment is more industrialised. Water quality is a problem for some industrial processes and managers are aware of the problems of water shortages for production and are making some efforts to reuse and conserve water. Industrial returns to water are in marked contrast to those of agriculture both in the provision of livelihoods and in economic terms. However this is not without cost. The growth in industrial pollution is essentially unchecked and attitudes towards it are rather ambivalent. Although there is an awareness of pollution problems there is very little action to rectify them and some may even use them as a means of land acquisition. Certainly specific manufacturers allow pollution to become an externality.

## Domestic Water Supplies

A greater proportion of people in the Ta'iz governorate live in a rural setting than the national average and those people are worse served by water and sanitation than the rest of the country. In contrast the urban population is served better than the national norm. In most rural areas water is fetched by women and/or donkey and consumption is very low at around 20 lcd. In some areas where there are adequate water resources and people live close together enough to make a piped network economically feasible the private sector can and does provide adequate water. A major problem facing piped schemes whether private or public is the supply of spare parts. Another problem in "community-run" schemes is the potential for corruption, much depending on the quality of the local shayx.

In the city of Ta'iz, there are very distinctive differences between the poor and the wealthy. The differences not only are seen in the more obvious forms of jobs and income, possessions, housing and level of education, but also in the water-related activities. Hydraulic differences are accentuated in the high water stress situation exhibited in Ta'iz. The wealthy have more water using facilities and the poor, especially children of poorer families, spend more time collecting water (Figures 3.6 and 3.7). The rich can better afford tankered water and the poor are more dependent on queuing for free water from standpipes (Figure 3.5). The wealthy were much more cushioned from the affects of water shortage and could afford the luxury to place water quality higher on their priorities than poorer people.

The quality of piped and tankered water is below WHO drinking water guidelines for over 95% of the households. The quantity provided by the public supplier is below those standards (Figure 3.8) and the total from all sources consumed by the "average individual" is at or around that limit. Water stress felt through severe shortage and declining quality results in more time spent collecting water. Although less time is spent using it, that time is concentrated in bursts of hydraulic activity particularly in washing clothes on the rare occasions that water comes out of the tap (Box 2, chapter 4). With water from the public utility being relatively cheap, the price for that source did not figure as important to consumers as water quality. Although the city's inhabitants are conditioned to adapt to water shortage those with a single, monopoly supplier, such as the public utility, are more vulnerable to water stress. The differentials described in this section provide the hallmarks of the Ta'iz urban hydraulic society.

*Water Markets*

Private piped supplies are a very small (1%) and declining portion of water supplies in Ta'iz. They are a monument to upstream resource capture, lack of monitoring, lack of regulation, commercially unreal administration, and community failure in allocating equitably.

The failure of the public utility to provide enough water or water of adequate quality for drinking has spawned private sector initiatives. Wells are the source of the private sector and run at a loss, particularly those within the city which are effectively subsidised by mosques. Water tankers provide an inequitable, expensive, alternative source of poor quality water to the public utility. The loss of economies of scale in the tanker delivery method is the "flip side" of being a good job provider. Irregularities in the quality of water provided by tankers beg enforced regulation of standards.

The drinking water treatment companies have proliferated as the quality of water provided by the public utility has declined. Although a reasonable price is charged, the proportion of income spent by most of the population on water is well above the World Bank recommended level of 2% of household GDP, and for the poor is above the maximum of 5% (Figure 3.11). Most household expenditure on water is for drinking water.

NWSA supply water for 10% of the time, at 18-30 day intervals. Although a block tariff is used by the NWSA the wealthy probably gain access to greater quantities of this government-subsidised source of water. Inconsistencies in the evaluation of public utility water provision make this observation uncertain, however, it may be stated that the NWSA water supply is inadequate in terms of quality, quantity and reliability, as well as being inequitable, inefficient, and financially unviable.

It must be asked whether the opportunity cost of queuing in the free water market is equitable and whether "free" water should be exempt from water quality standards simply because it is free. Despite the failings of the private sector in maintaining quality standards and providing for the poor, the private suppliers have proved themselves adaptable to water stress and have provided water when the public utility has failed – but at a price.

# 4 Discussion:
# Anatomy of a Water Crisis

In this chapter, development theory is tested against the empirically derived physical and socio-economic realities of the Ta'iz situation. These realities were presented in chapters two and three. Where appropriate, new data are presented in this chapter to illustrate the points made.

The evidence given in chapter three suggests that the water shortage in Ta'iz not only had a significant impact on the social habits of the population, but that human activity largely contributed to the crisis in the first place. Water can 'flow uphill to money' (Reisner, 1986), and this chapter begins with a summary of the economic context which underlies the social adaptations to the water shortage. Reisner continued by suggesting that 'water flows uphill to money and power'. Political ecology seeks to unravel the 'interactions of different actors pursuing their quite distinctive aims and interests' (Bryant and Bailey, 1997; 21) and the second section comprises an actor-oriented analysis (ibid;24) of the actions of the powerful in determining water allocation in the Ta'iz area over the past thirty years. Within Bryant and Bailey's framework of political ecology (1997; 21-24) the Ta'iz situation is problem specific – water shortage (reflected in declining groundwater levels and water quality) and regionally specific – that is, Ta'iz and the surrounding Upper Wadi Rasyan area.

Coase suggests that the legal system under which a market operates largely determines whether the transaction costs of making an exchange can be covered (ibid; 717). In particular he identifies 'clarity in the law', 'an appropriate system of property rights' and whether 'they are enforced' (ibid; 718) as necessary if trade is to be facilitated. The third section of this chapter examines the existing legal framework in Yemen regarding water transfers from the specific standpoint of these three issues. Coase (1992; 714) also asserts that appropriate institutions are needed to govern the process of exchange in a market economy. Water transfers in Yemen are no exception. The fourth section examines the existing institutions in the context of their appropriateness to facilitating water transfers.

An adequate examination of the origin of the aims and interests of the political actors that operate within, and manipulate, the legal and institutional frameworks, is beyond the scope of this study. However, a few observations regarding some prevailing belief systems concerning water allocation and the implications for development are given in the fifth section.

Finally, the problems of water resources management and efficient water allocation in Ta'iz have not occurred in isolation, but rather have attracted a considerable amount of interest and money from the development community. The final section considers the concept specific aspects (Bryant and Bailey, 1997; 21-24) of economic progress, equity provision and environmental protection in the water sector in the contexts of both current debates in the development / donor communities and their involvement in Ta'iz.

## The Role and Limitations of Demand Management in Ta'iz and the Virtual Water Solution

During the 1990s the international hydraulic community has progressed in its awareness of appropriate measures needed to alleviate water scarcity towards making the water go further rather than simply trying to find more water. Those measures are summarised under the umbrella term "water demand management" (WDM).

The steps in awareness which take us towards formulating a demand management policy may be summarised as:

1.　　Water has a value.
2.　　Water is an economic good.
3.　　Water should "gravitate" to the highest value use by:
　　　a) Productive Efficiency (promotes the reduction of waste and the more efficient technical use of water).
　　　b) Allocative Efficiency (promotes intersectoral water transfer).
4.　　Allocatively Efficient water transfers are facilitated locally via water markets.
5.　　On a larger scale facilitated via virtual water (international water transfer).

Although the hydraulic actors in the city of Ta'iz, Yemen are mostly remote from the international discourse, to an extent, this progression has occurred *ipso facto*. It is suggested that, as a consequence of the needs and economic realities even if not by conscious awareness, the implementation of WDM principles is directly promoted by increasing water scarcity and that more water scarce regions are likely to have progressed further in their implementation than many water-rich areas.

*Water has a Value vs. Water as an Economic Good*

That water has a value and that it should be treated as an economic good appear to be two sides of the same coin. The first, water having a value, is related to the benefits accruing from water. The second, water being treated as an economic good is related to the costs incurred in accessing water. They form separate steps in the awareness process, however, simply because there seems to be a time gap, if not decision-making gap between appreciating the value of water and being prepared to pay for it. Although many can appreciate some of the user aspects of water (drinking, irrigating, manufacturing, swimming etc.), the full value of water including its future value and value to non-users identified by Burrill (1998) is impossible to quantify. The UK motor industry is said to provide four times as many jobs in related provision of services needed to maintain it as there are people directly employed by it. This "trickle-down" phenomenon is also observed in the water supply of Ta'iz where its value in terms of livelihood provision to those who trade, transport and treat it (as described in chapter three) is considerable. The greater the losses of economies of scale in the private sector the more are the benefits in livelihood provision.

When it comes to paying for the water, the current development ethos informs us that the economic costs of its provision should be recovered. However, the awareness that water is not free, and, more importantly, should not be free is slow to permeate any society. Ta'iz provides some useful examples of different aspects of the cost of water.

*Marginal costs* The public piped water utility is failing to collect adequate revenues to even cover half of the running costs from an average charge of only 0.17 $/m$^3$. Average incremental costs of alternative supplies currently being considered range from 0.5 to 1$/m$^3$ from conventional potential groundwater sources to 2.5$/m$^3$ for desalinated seawater (Handley, 1999b). These estimates exclude the provision of sanitation and also salaries. In rough terms current tariffs are possibly around a quarter of the marginal cost. Although the public utility shows little indication of progress in treating water as an economic good, the expenditure by the public on water when all sources are considered (Figure 3.11) suggests otherwise. It is considered that the price paid by households reflects water's necessity status which forces a willingess-to-pay above and beyond capacity-to-afford and also forces water to become an economic good in consumer awareness. Pricing by the public utility, and the consumer attitude to the utility on the other hand suggest that water is a public right rather than an economic good. This problem is perceived by the author to be related to prevailing belief systems and also to the institutional and political context. These aspects are discussed in later sections. If marginal cost recovery by the public utility is to occur then significant changes in the utility will be needed, though consumer capacity-to-afford may not be enough.

In the agricultural sector, the irrigators of nearby Wadi Ad Dabbaab who also supply tankers were asked whether they would sell bulk supplies to the city of Ta'iz. Although they said they would, all made the proviso that in the case of scarcity irrigation would come first. This is despite the fact that returns to water from supplying tankers are better than from irrigation, or rather are not as bad. Farmers do not face up to the true marginal cost of irrigation, but then nor does the World Bank (Moench, 1997; 12). The reason the farmers do not is not that they are poor money managers. The 1980s remittances boom which paid for the wells blinded them to the real costs, and subsidies on pumps, irrigation equipment and diesel since has tightened the blindfold. The few who still have enough capital / power to drill are spurred on by the increasing scarcity of underground water, the exploitation of which still remains unchecked by regulation.

*Opportunity costs* Two examples are cited as evidencing the preparedness of consumers to pay significant opportunity costs. One is the queuing of people for water. The social benefits for rural women waiting their turn at the well (Ansell, 1980, Maclagan, 1995) may significantly offset the opportunity cost of being able to spend their time doing something else. The contrast between wealthier and poorer peoples' access to free water by children or men queuing and tankers by payment in the urban environment (Figure 3.5) suggests there is a household evaluation of opportunity cost in which time and money are exchanged.

The second example is reflected in the effort made by industrialists to ensure a reliable supply of water. The major industrialists interviewed all consider water provision in the context of days of production lost due to insufficient water supply. They therefore make opportunity cost driven decisions and are perhaps more in a position to do so because water is still a relatively small part of their total inputs budget.

*Transaction costs* The very existence of the water markets in Ta'iz indicates that the transaction costs incurred in these markets have been overcome. However whether the transaction costs of bulk rural – urban transfers of water are met is another matter. The transactional cost of the World Bank "experiment" in negotiating water transfers from Habir to Ta'iz has cost around $700,000 for a supply unlikely to provide more than 20 lit/sec to the city. This amount does not include the $8M compensation being paid nor the exploration costs ($2.5M). If the scheme lasts thirty years, the negotiation costs will be around 4cents/m³. If the exploration costs are included and a more typical ten-year life is assumed, then the transaction costs are 50cents/m³, that is, around three times the price that the same water is sold for to the city consumer by the public utility.

*Social costs* Dellapenna (1995; 155) points out the potential social costs involved in selling water and cites the potential loss of tax revenues to a community as an example. This cost might be considered a form of opportunity cost, and would certainly be a factor if, for instance, water sales to the city from agricultural use resulted in a decline in crop production, and hence a drop in zakaat due from the crops. Under-collection of zakaat and non-transparent returns to the community mitigate against the significance of this cost, and in any case, under Yemeni customs, irrigated crops are taxed at a lower rate than rainfed crops (a wrong economic signal to the irrigator). The little tax lost from the agricultural sector could be recovered through the water use activity in the sector the water is transferred to (industry more so than domestic use). However, the means of returning benefits accruing via taxes on irrigated products will have been lost to the water source communities through the water transfer.

*Intrasectoral Productive Efficiency Measures*

Discussions of transfers of water within the agricultural sector in the development literature tend to focus on irrigation methods and crop varieties and there is certainly scope for such measures in the Upper Rasyan catchment. The switch by farmers from coffee to qat production may be construed as one of the latter measures, although those who think qat uses as much water would contest this (see Weir, 1985, Al-Hamdi, 1998, Heidbrink, 1994). However, three points are much more salient than the application of such measures. Firstly, regarding irrigation methods, savings in water as a consequence of switching to a more efficient irrigation system will mainly reduce "losses" which would otherwise have recharged the aquifer. Savings from improved irrigation methods will be mainly in the energy (usually diesel) expended in pumping the water rather than in protection of the water resource itself. Secondly, because there is more potentially cultivable land than there is water to irrigate it, any savings in water will only result in the irrigation of a larger area rather than a reduction in water use, that is, water is the constraint, not land (Bromley, 1986; 593). This was clearly demonstrated in farmer decision making in 'amran (Handley, 1996b) and leads to the third point, that the principle of the "tragedy of the common" groundwater is a ubiquitous driving force. As one farmer in 'amran put it 'if I were to try to reduce my water consumption' (specifically by installing drip irrigation in this instance) 'my neighbour would get a larger slice of the resource without the expense I, the water-efficient farmer, have incurred' (ibid.).

Traditional rainwater harvesting methods transfer water efficiently from land with very poor cropping potential to marginal land where but for the extra run-on cropping would not be possible. Another ancient, indigenous productively efficient form of water use is rainwater harvesting from land and roofs in the rural and urban areas for domestic and animal use. A more recent development in the implementation of productive efficiency measures is industrial water recycling and industrialists are currently exploring further steps in this area. The reuse of domestic waste water by agriculture has been poorly managed with degradation of the environment as a consequence. Waste water treatment is needed to make agricultural use a viable option, but at around $0.6\$/m^3$ the costs are probably prohibitive.

*Intersectoral Allocatively Efficient Water Transfers via Markets*

With six discernible urban water markets and the reality of intersectoral reallocation of water resources, awareness and implementation of allocative efficiency measures in Ta'iz are surely well ahead of "Western" counterparts. However, the volume of water being transferred by agreement from irrigated areas with low (or rather negative) returns to water to domestic and industrial users who pay economically realistic rates is limited to around $2.5Mm^3/yr$ (shown underlined in Table 4.1).

Despite Falkenmark and Lundqvist's assertion that 'reliance on market mechanisms is often less than realistic in the 3rd world' (1995) and Dellapenna's claim that (1995; 153) markets in water have never actually played much of a role, the very opposite has occurred in Ta'iz. Over 75% of the population have depended on the private market for drinking water costing around 20$/m³ and up to 1/3 on tanker supplies for domestic supplies costing around 2$/m³. In times of even more severe shortage participation in these markets increases further.

**Table 4.1 Summary of water users and intersectoral transfers**

|  | Quantity Mm³/yr | Price $/m³ |
|---|---|---|
| Rainfed and Stream-fed Agriculture | 103 | Free |
| Irrigated Agriculture | 30 | 0.08 – 0.16 |
| Rural Domestic Use | 2.5 | "Free" ie. 20 hours / household /m³ |
| Urban Domestic Use | 2.7 | 0.17 |
| Industrial and Commercial Use | 2.5 | 0.17 – 2.0[a] |
| Water Transfer: Rural Agriculture to Urban Domestic via tankers | 0.5 | 2.0 |
| Water Transfer: Rural Agriculture to Urban Domestic via NWSA Supplies (quantity pumped) | 5 | 0.17 |
| Water Transfer: Rural Agriculture to Urban Industrial via tankers and pipelines | 2 | 0.5 – 2.0[b] |
| Water Transfer: Virtual Water imported from overseas to the area in grains | 100[c] | 0.11[c] |

[a] NWSA / private tanker source.
[b] Own tanker / private tanker source.
[c] See text.

It might be asked why water markets have become so prevalent in Ta'iz and what has facilitated the intersectoral allocative transfers that are so difficult to establish in other places (Allan, 2000). A shortage of supply from the public utility, coupled with a decline in its quality and accentuated as the water shortage has deepened, has spawned the tanker and drinking water markets. The shortage has become so extreme that the physical constraint of moving bulky general domestic use water by tanker, or the local treatment of drinking water with all the loss of economies of scale involved, can still be funded at least by up to one-third and three-quarters of the community respectively. The high industrial returns to water combined with the significant political power of the industrialists has enabled them to meet their basic water needs through transfers from the agricultural sector.

In the view of Ward and Moench (Unpubl.) and this author the Ta'iz water markets are likely to expand. However, as noted in the instance of Yarim (Handley, 1997a), the limit to expansion is not likely to be the availability of tankers or pipelines but of the resource, and water or its food equivalent has to be sought from outside the surface water catchment.

## Suppressed Demand Management

It seems rather hypocritical for a Western expert to espouse the merits of demand management to a Yemeni, at least to a Ta'izi, and then return home to her "Western" hydraulic lifestyle whilst the Yemeni goes home to his 25-30 l/c/d lifestyle. Suppressed demand is defined in the Ta'iz instance as the absence of choice to increase one's consumption. Essentially that choice is not open to most households, they simply have to make do with what they can get. A similar situation prevails in the rural areas where a lack of plumbing infrastructure adds to the problem of inadequate supply and limits consumption to 25 l/c/d or less. The adaptive capacity of Yemeni households in suppressed demand conditions to cope with the water stress is remarkable (Box 2).

---

**Box 2: Ta'iz Water – The Reality**   (author quoted in World Bank, 1997)
'When the mains water finally arrives, all social engagements are cancelled and the mother and daughters will work from 6:00 am to 12:00 p.m. for the one or two days the water is connected. The washing has accumulated into a huge pile, some clothes are being worn a second time over and there are no clean clothes or bedding left in the house. The next day after the water stops is drying day. There is usually not enough room, so clothes are often draped on the roof over any object available – reinforcement bars, water tanks, etc. The next event is ironing. The whole cycle is about 4 – 5 days of constant water related activity by all female members of the household.'

*From the peak of the water crisis in August 1995 when water was coming once every 40 days.*

---

It is difficult to envisage, and unrealistic to expect, much improvement in end-user efficiency. If more water were to become available, then stronger demand management measures would be likely to be needed. Although suppressed demand results in, or equates with, reduced demand, it is neither a solution to inefficient use, nor a substitute for demand management.

Despite the adaptation to shortage of individual households there is a need for a similar adaptive capacity in the Yemeni institutions and political economy to redress the factors which have led to the shortages that have developed under their supervision. These issues are discussed below in relation to their political and institutional contexts.

### International Allocatively Efficient Water Transfers via Virtual Water

The importance of imports of "virtual water" embedded in grain, to the Middle East in particular, has been described by Allan (1998). Yemen is no exception, and has steadily increased its import of wheat and flour since the mid-1970s (Figure 4.1). In the period 1975 to 1996, total cereals imports have increased from 0.5M tonnes to 2.5M tonnes whilst domestic production has remained relatively static at around 0.5M tonnes (World Bank, 1998a; Figure 2). That the dependency of this, one time "bread basket of Arabia" on the international staples market began at exactly the same time as the migration of workers to Saudi Arabia and the Gulf in search of remittances is no coincidence. However, the return of a large portion of those workers in the wake of the Gulf War has not resulted in a decline in the import of grains. Local production of grains has not kept pace with the huge increase in population over the last 25 years, but rather, significant tracts of land that used to produce sorghum are now used for qat cultivation.

Ta'iz is no exception to the national picture. The household survey indicated that at least 91% of the population of the city purchased wheat or flour in bulk (50kg sacks) and the average household consumption was 24 sacks per year.

Despite a preference for home-grown sorghum only 15% of households brought an average of 3.7 50kg sacks per year from the village, that is, one fortieth of the imported grain consumption. Ta'iz city population represents 2.7% of the national total and their wheat and flour consumption as indicated by the survey is almost exactly in the same proportion. If the city consumes an amount of wheat and flour close to the national average, then rural areas must also (although government subsidies provide incentives for leakages e.g. via smuggling).

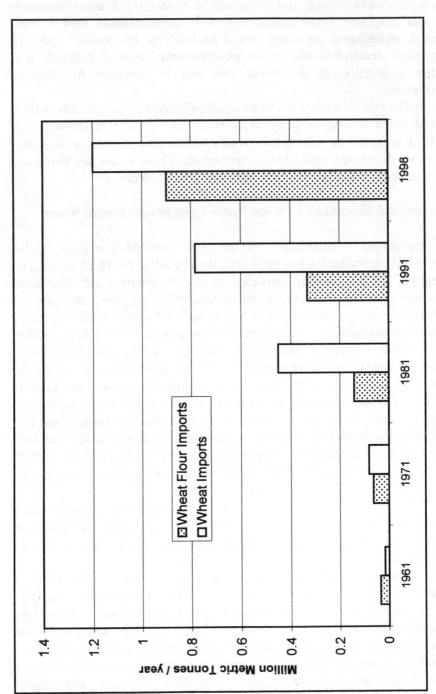

**Figure 4.1 Virtual water imported to Yemen** *Source:* FAO Agrostat.

The equivalent quantity and price of virtual water indicated in Table 4.1 assume 1000 tonnes of water per tonne of wheat produced, that 4% of the national total reaches Ta'iz and the surrounding catchment of Upper Wadi Rasyan, and assumes the April 1998 street price of 770YR (5.9$)/50kg sack of wheat.

That significant supplies of imported wheat and flour are being taken from the city depots to the rural areas, would explain the sacks seen on nearly every Toyota Landcruiser returning from city to village. Virtual water in Yemen has received two huge subsidies. Firstly the subsidy of the exporters paid to their farmers and secondly Yemeni subsidies to its importers. The result is the provision of virtual water at a price as cheap as the cheapest local source of water, that used in irrigation, and in amounts as large as the water used by local rainfed agriculture and almost as large as the total water economy (Table 4.1).

*Some Equity Issues*

Although the progression towards reallocating scarce water resources to higher value uses may make good economic sense, significant equity issues are raised. Among these are:

1.  Compensation and the issue of protecting third party interests. That is, all those who had livelihoods dependent on the water sold in the selling community, not just the well owners.
2.  The interests of those who cannot afford to purchase water from the markets, or who have to pay a significant portion of their income on water in order to survive. That is, 2/3 of the urban community who cannot afford tankers when the piped water runs out and 1/4 of those who cannot afford treated drinking water when the piped water is of inadequate quality.
3.  Those who miss out on the 'predetermining allocative logic' of a piped supply, (Falkenmark and Lundqvist,1995;214), that is, those on the ever-expanding accretive edge of the city and/or those who cannot afford to connect.
4.  Those who access less from the effectively subsidised piped system, that is, those without gardens and without many water using facilities available to the wealthy.
5.  Those paying the externality of pollution, that is, those on the receiving end of domestic and industrial waste water and those without sewerage.

These people represent the losers in the water providing processes described in this section. Although the issues involved are beyond the scope of this section, and are reserved for later discussion, the inequitable consequences of water allocation must be noted.

*Summary*

It would appear that there is a difference in the application of WDM principles if not in awareness of them in the various water-use sectors of Ta'iz. In the irrigation sector where water, as opposed to land, is the constraint, productive efficiency demand management measures have little impact due to wider "structural" issues of the price of diesel fuel, land availability and the commonality of groundwater. Besides the irrigation sector, those who do not act as though water were an economic good include the public domestic supply utility, and, it is suggested, the "Western" development fraternity. In contrast, high levels of productive efficiency in water use by rainwater harvesting have been practised since the habitation of Arabia Felix began.

Also in contrast to the irrigation sector and the public utility, in the domestic and industrial sectors, where capital is a greater constraint, efficiency measures which come under the umbrella of demand management are seen to be applied and to work. Urban domestic and industrial users pay significant, though not full opportunity and transaction costs. Whether the marginal costs of domestic supplies will be met is perhaps more of an institutional issue than an economic one. Certainly domestic drinking water and industrial process water marginal costs are being met.

Those who do regard water as an economic good include the industrialists, those who trade, transport and treat it, and those who have to buy it to meet their basic drinking needs. The split between those who consider that Water Is an Economic Resource (WIER) and those that do not (WINER, Allan, 2000 terminology), is related to small volume high quality drinking water in the former camp versus large volume lower quality domestic / irrigation water users in the latter. The main exception to this rule is the industrialists who need quite large quantities of reasonable quality water and are ready to pay WIER prices. The reason that they can be the exception is the far greater returns they achieve from their water use when compared with other users (Table 3.12). The industrialists and the water drinkers of Ta'iz seem to believe that Water Attenuation Transforms it into an Economic Resource ie they are WATER users.

The fact that individual households and businesses have turned to the highly need-adaptive water market is essentially due to the failure of the public utility to provide adequate supplies. The adaptivity of those households and businesses to the water shortages reflects the suppressed demand conditions that have led to the proliferation of alternative water sources in adaptive markets. The growth of the markets has also been helped by the prevailing environment of political interests and legal pluralism that are examined in later sections.

Local water markets are, however, dwarfed by the scale of involvement in the international water market. The amount of virtual water imported from overseas in the form of grain underlines Yemen's dependence on World Water.

Perhaps the most significant problem in both the intra- or intersectoral reallocation of water but particularly the latter is one of equity. This is particularly so for those in the buying community who cannot afford it and those in the selling community who are not direct recipients of the benefits of the sale.

This section has demonstrated that there are significant economic incentives to promote the transfer of water from irrigation to domestic and industrial sectors, and food needs are met by the import of virtual water. Water transfers, however, do not occur on an adequate scale to meet even the modest current demand and reliance on virtual water is ignored. The reasons for these inconsistencies are explored in the next section.

## Hydropolitical Reality: Economic Sense vs. Political Expedience

So far in this chapter, we have discussed some of the economic incentives and constraints to the efficient allocation of water in the Ta'iz region. However, in the water use and allocation business 'economics are an illusion; politics are real' Reisner (1986). Allan also notes that the pursuit of productive efficiency measures are politically acceptable whereas the much more effective reallocative solutions to water scarcity advocated by economists and scientists are politically too costly (2000; chapter 4). The previous section demonstrated the economic sense of the reallocative hypothesis in the Ta'iz instance. This section explores whether the politically possible prevails at the expense of the economically sound.

Aptly describing the hydropolitics of Ta'iz, Migdal (1988) asserts that the accommodation between the state and other interest groups, their struggles and manoeuvres constitute the "real politics" of the East. In this process, Hajer (1995; 39) emphasises the importance of the "sub-politics" of the concealed, individualised, micro-powers as opposed to the acts of a single sovereign. This section seeks to explore the "real politics" and "sub-politics" of some of the water issues affecting Ta'iz. In order to do this the national historical setting is very briefly examined. Then, based on the area of Al Hayma / Habir which has played so prominent a role in the water supply of Ta'iz and its demise, the following questions are asked:

1. Who are the actors?
2. How have they influenced, or controlled water allocation?
3. Whether gains and losses in the resource capture game reflect power asymmetries?

The political influence of some wider regional, national and international issues on local water allocation are then discussed and the capacity to adapt to water scarcity by those involved is examined in conclusion.

## *The Political Map: Historical Roots of "North" and "South" Yemen*

The history of Yemen can be characterised by the recurring theme of North – South opposition. Dresch (1989;11) states:

> the northern tribes ... from the end of the ninth century AD to 1962 were associated with a succession of Zaydi (Shi'ite) Imams ... The southern mountains and the Tihamah have been predominantly Shafi'i (Sunni) for almost as long, being ruled by a succession of more or less powerful states and sometimes raided or dominated by the northerners.

Ta'iz lies within these southern mountains and has had its fair share of northern domination. Phases of northern domination include those of 1636 (to expel Turks) and 1823 (to avoid drought). Each time the northerners moved south permanently they ceased to be tribesmen and became landlords or peasants like the people around them (ibid; 13) in a similar manner to the Bedouin raiders in Oman (Wilkinson; 1977). Domination has been characterised by the extraction of taxes 'by northern tribesmen attaching themselves to government officials in the south without any appointment by the state .. and stayed there permanently' (Dresch; 229).

The overthrow of the Imamate in 1962 and the establishment of 'the Republic constitutes an obvious fracture with the past with an equally obvious continuation of the tribal forms within it' (ibid; 236). That continuation, in the opinion of many Ta'izis, includes the extraction of taxes from the region, particularly from the fledgling industries. They call Ta'iz "Al baqara al huluub" – the milked cow. Taxes are widely perceived as the means by which the northern tribally-based government officials obtain their "layla 'alawi" Toyota Landcruiser (or even more curvy "Monika" – ref. Clinton, 1999 model). The north – south divide appears as strong as ever and, despite being part of North Yemen prior to unification, Ta'iz places its affiliations firmly in the south.

The demise of Russian communism led to the end of effective sponsorship of South Yemen, and unification of the two Yemen's followed in 1990. Despite the honeymoon ending with the civil war in 1994 and the abolition of the southern-led socialist ('ishtiraaki) party, exposure to socialism has at least left some notion that equity, instead of divinely appointed poverty, is an option. Indeed, some water projects were undertaken in the Ta'iz region under a socialist banner. Since 1994 the resurgence of more overtly Islamic interests in the form of the 'islah opposition party has occupied much of the ruling Mu'tamar party's energies. Border bickering with Saudi and keeping the south under the thumb occupies most of the rest.

*The Case History of Al Hayma – Habir: the Events: 1976-1995*

During the fieldwork and exploratory drilling in wadi Al Hayma from 1976 to commissioning in 1982-1983, several misconceptions seemed to have arisen. Some locals thought that NWSA were only to drill seven wells, and now there are over thirty (Ward and Moench, unpubl.). The farmers were told that the pumping of deeper wells drilled for NWSA would not affect their shallow dug wells because there was an aquiclude in between. When four years abstraction resulted in cessation of the perennial flow of the wadi and the dug wells had largely dried up, the absence of an aquiclude was all too apparent. $10M compensation for loss of crops had been promised (Tipton and Kalmbach, 1979) but none had been received. Local farmers appear to have been at least uninformed or misinformed regarding the implications the NWSA abstractions would have on them. The shayx of Lower Al Hayma, agreed to the drilling and managed to get three deeper, drilled wells out of the deal. They continue to supply his successors' lucious qat plantation via open channel irrigation to this day in the midst of an otherwise once fertile, now barren, valley (Photo 2).

Prior to the emergency drilling campaign in Al Hayma in 1987 the NWSA wells reached a maximum depth of 100 to 120m. The new wells included in the campaign were up to 500m deep, but government did not allow local farmers to deepen their wells. [The shayx of Lower Al Hayma managed to obtain one of these deeper wells.] The locals then stopped the NWSA drilling. The army was brought in and allegedly took the school children hostages in the school (for a day) whilst the local men went into the mountains with their weapons. Five shayxs were imprisoned by political security in Ta'iz for five days. A first attempt to resolve the dispute by the head of the co-operatives at mudiiriyya (now nahiya) level, was rejected by the army. The local people then approached the governor and contacts in San'a requesting them to mediate. The Minister of Oil came down and succeeded in obtaining the release of the five shayxs on condition that they signed an agreement to not hinder the progress of drilling. The sons of the shayxs guarded the drilling rig (part of necessary feigned commitment to the government position and to relieve the situation).

After it became known that the Habir area was of interest for exploratory drilling, in June 1992 some locals disconnected one of the government wells in the Xazaja area immediately to the south. It has since been claimed that this action was supported by the socialists who were seeking a local grievance from which to generate political mileage. The government responded by placing 20 trucks of soldiers from the Ta'iz military camp on stand by. Helicopters were sent on reconnaissance missions. Some socialists (Al 'amba) were imprisoned. The president's man in Ta'iz, the deputy head of security, managed to get the electricity and water reconnected and a co-operation agreement from the locals was obtained. The agreement was forced under his threat 'if you do anything more, I am not responsible for the consequences' and the troops were not sent in.

The interests of the Dhi Sufal area had been protected their powerful shayx, who was Minister of Agriculture at the time. Through his efforts, a well drilled previously by military strength as part of the exploratory campaign for Ta'iz was connected to supply Al Qaa'ida instead. Exploratory drilling in Habir again met with local resistance in Dec 1993. All the main shuyuux in the area were invited up to San'a to hear the president's immortal words 'you will co-operate with the drilling', "imma bil 'urf aw bil 'unf' – 'either by custom (that is, gentleman's agreement) or by violence'. In June 1995 the Dhi Sufal shayx convinced the government that the five existing wells in the Dhi Sufal-Al Qaa'ida valley were the maximum number it was possible to equitably operate and succeeded in getting drilling attention moved to Habir.

In April 1995 a rig arrived to start drilling six exploratory wells in Habir (three in east and three in west Habir). The locals threw the drillers off site in the same month using petrol torches. Three different visits were made to the area by up to four ministers at once to discuss the problem. Visitors included the Ta'iz and Ibb governors, the minister of agriculture, the minister of water and electricity and the minister of civil services. All the shayxs involved were then summoned to Ibb and asked to promise to protect the drillers. They complained that they could not make promises regarding their irate peoples' behaviour, especially considering their people's mistrust of the government, reinforced by the Al Hayma disaster. Three shayxs were imprisoned in Ibb and four in Dhi Sufal until they agreed to sign an agreement of co-operation. Given the alternative of remaining in prison whilst the government used force to drill the wells they agreed to sign. The agreement incorporated a compensation package.

Hostilities flared up again in January 1996 when the government (NWSA) failed to deliver part of the compensation. The drilling was stopped, the governor responded by sending the troops in, so the villagers sent women and children to stop the troops. The women threw stones and tried to take the weapons from the soldiers and two women were seriously injured when one of the soldiers opened fire. Fortunately, further escalation was prevented when the local shayxs and the governor of Ibb intervened. The subsequent involvement in the area by the World Bank through the TWSSP (see institution section, below) has witnessed various extensions to the compensation package including the provision of water supplies to the nearby villages. This accounts for around 25% of the total yield of the three wells connected so far. With 50% of the remainder to be lost through leakages, a costly compensation package and considerable civil strife in hindsight it must be questioned whether the whole exercise was worth it.

SURDU had been conducting a dam feasibility study in Wadi Warazan, and in 1995 a rig went there to drill a site investigation borehole. Specifically because of the Al Hayma experience, the local shayxs were suspicious of any drilling, suspecting it was for the purpose of abstracting water for Ta'iz. On a pre-arranged day, 5000 armed men converged on the drill site and halted the drilling. The governor of Ta'iz came to negotiate and left again having agreed with the locals that no drilling for Ta'iz supplies would take place and no dam would be constructed.

As a final incident for consideration, the handling of the Ta'iz emergency drilling within the city during 1995 is worth noting. Although technically poorly planned and executed, with resulting poor yields, from a political perspective it was more successful. The governor pushed the project to help alleviate the water crisis (or was it to increase in his popularity in the city, or both?). Having raised finance for the drilling from local businesses, the major outstanding cost was the pumps. The governor requested the money from central government under threat of his resignation if it was not forthcoming. His bluff was called, he resigned, the money appeared and his resignation was refused (Al 'Ayyam; 7/6/95, Yemen Times; 12/6/95).

## Some Observations

Although many observations could be drawn from these events some are particularly prominent:

*Strong society – weak state* The Al Hayma – Habir incidents have demonstrated how state intervention can go amiss locally, and that struggles over environmental and socio-economic issues at the local micro-level can have a momentous impact on the state and its goal of predominance (Bryant; 1998, Migdal; 1988). Compared with many countries the ability of local communities to resist the government imposition of wells which would "steal their water" demonstrates Yemen as having a strong society and a weak state (Migdal, 1988). The phrase 'compared with many countries' is used on purpose because many Yemenis believe that even the consideration of using military intervention to drill, let alone allowing water to be abstracted for the city could never have occurred in the North of Yemen. They equate this high-handed government position to be tenable only because the tribes are not as strong in the south compared with the north where society is even stronger and the state even weaker.

*Principal – agent problems and knowledge asymmetries* The actions of the Al Hayma shayx clearly demonstrated:

1. The potential for principal – agent abuses which became worse as the as the arena grew larger than the immediate locality. The agent (shayx) was given scope for opportunistic behaviour and received the welfare (water) of the principal (the local farmers).
2. How central institutions (NWSA) could offer an escape route from the demands of traditional institutions (the shayx role), undermining communal responsibilities.
3. The cause-effect relationship of the NWSA drilling on existing wells was not perceived by the man-in-the-field, in fact he was lied to. The perpetration of this particular principal (farmers) – agent (NWSA) abuse was only possible because of the knowledge asymmetry between them.

4.      The decision to exclude people from using the resource did not result from exhaustion of the resource, but was consciously made beforehand (unlike Bromley, 1986; 594).

The net result of these principal – agent abuses was that the hand dug wells of the many ran dry whilst the deep drilled wells of the few, particularly the agent, continue to yield to this day. This is a clear example of the control of one actor over the environment of others (Bryant, 1993; 11) and of resource capture that directly reflects power asymmetries. The declining influence of the agent's descendants over the same principals because of past abuses indicates an underlying "democracy" in the shayx system.

*Emblems and the political process*   Hajer's concepts (1995) of an environmental emblem or rallying point and of the long process from local environmental problem to political issue status are apparent in the history of Al Hayma. The declining water level in Al Hayma was both an environmental and socio-economic problem. However, since an urban population with its tarmac, capital biases (Chambers, 1983) has a stronger political voice (Falkenmark and Lundqvist, 1995), the "problem" surfaced as a drinking water supply issue, although it was actually part of the same problem. In the end, it took politicians (governors and ministers) to recognise the political mileage in it and turn a problem into an issue. One of those politicians (the governor of Ta'iz) saw the window of opportunity and forced the hand of the president in getting the pumps paid for. A sobering aspect for the windows-of-opportunity adherents is that the very fact that a crisis creates the window also means that the decisions taken are crisis-driven. As such they may result in only soothing the symptoms and have unpredictable, knee-jerk, non-optimum outcomes, as occurred, it is suggested, in the case of the governor's emergency drilling. Of course, to maintain the knowledge asymmetries, the general public are not informed of the failures (Yemen Times 21/8/95; front page). The rise of Al Hayma to emblem status is confirmed every time a drilling rig (the most evocative emblem, incidentally) turns up on site. Each time a rig appears local leaders specifically mention Al Hayma as the reason the rig must depart.

*Political "Riyalities": Self-interest in Regional, National and International Contexts*

*Equitable local water transfers: compensation vs. opportunism*   The distinction between the compensation of water source-area communities and opportunism on their part, particularly by the well owners, is slight.

One task in establishing equitable water transfers is ensuring that compensation reaches the whole hydraulic community in rural source areas, that is, including those without wells but dependent on irrigated agriculture for their livelihoods (termed third parties). This is the greatest challenge to the World Bank efforts at agreeing water transfers from Habir to Ta'iz. The track record of government water transfer methods, with the benefit of the very few and the misuse of knowledge, has engendered a basic mistrust which may only be overcome by erring on the generous side. Being over-generous may be labelled opportunism, but it needs to be over-generosity to all. Erring on the generous side appears to be the private sector example of Hayel Said's Soap and Ghee factory in providing electricity, piped water and jobs to those affected by their abstractions. In fact, considering water transfers throughout the region, the relatively smooth farmer – tanker / private pipeline transfers are in marked contrast to the central government example. The essential difference between them is that the former involves money transfer and the latter does not. A commitment to purchase water from farmers, so that water flows one way and money the other, would appear to be the way ahead if transfers are to be more like market transactions than diplomatic negotiations (Sax, 1994; 13). However, the Lower Al Hayma shayx is not likely to be the last political actor to benefit from government attempts to reallocate water at a level disproportionate to his original stake in the water.

*National issues: diesel prices and water law* It is a curious fact that a well owner can sell water to tankers or farmers for as little as $0.1\$/m^3$, whilst NWSA cannot cover basic running costs (exclusive of maintenance, depreciation or salaries) when charging twice that price. The main difference in their running costs is the price of power. In approximate terms it costs the farmer $2YR(0.015\$)/m^3/100m$ lift (1998 diesel price of $10YR(0.077\$)/lit$) compared with $9YR(0.07\$)$ for NWSA to lift one $m^3$ of water 100m by electricity. This difference in cost reflects the price of electricity to NWSA (charged at the maximum tariff) compared with the price of diesel to the farmer (charged at half the international level for oil exporters, Ward; 1998). If irrigators made the mistake of purchasing an electric pump instead of diesel, the increase in electricity prices over the past few years will have forced them to give up irrigating (Handley,1996b). Very few farmers made this mistake. The government is effectively subsidising the power costs of irrigators and over-charging NWSA, who have to pass on this cost to the urban consumer. A strange twist in the tale is that the ministry that sells such costly electricity to NWSA is the same ministry that pays for it – the Ministry of Electricity *and* Water! Central government power pricing subsidises the well owner to use over 90% of the country's water to grow, amongst other things, lucrative amphetamines, whilst charging over four times as much to provide a meagre 25 lcd of poor quality water for domestic use.

The cessation of perverse water pricing (Mollinga, 1998; 254) through diesel subsidies would contribute to improving local resource management by changing the net benefits of resource use (Steenbergen, 1996; 204). However, Yemen's stance on power pricing makes both good political and economic sense, since those in political power are the landowners. This reality has been apparent each time the government (under World Bank structural adjustment incentive) changes the price of oil products. Since the onset of structural adjustment, the major cost in oil products has been borne by petrol (affecting the man in the street) rather than diesel (affecting the ruling landowner class). When the Prime Minister increased the price of diesel in November 1995, the President appeased the landowner backlash by reducing the increase. Diesel price increases in October 1997 to 10YR (0.077$)/lit were an attempt at covering the real production cost of 28YR (0.22$)/lit (Al-Thawra 23/10/97) and sparked off road blockades by protesting farmers which resulted in three dead (Al Hayat 22/10/97).

> MP's rejected the rationale forwarded by the Minister of Oil and Minerals who maintained that the increase in the price of diesel was inevitable in view of the government's Economic Reform Programs. (The Daily Chew, 25/10/97)

but then, self-interest demands they reject it.

The same political "riyalities" underlie the failure of attempts at passing a water law and in particular the regulation of drilling and well spacing. Despite water laws being drafted on several occasions, they have never even reached parliament. Not only does Yemen have Migdal's problem of the strong society of tribal chiefs resisting the state (1988), those chiefs are the state (The President in Dresch; 1989; 7) acting in self, not state, interest. In Yemen's case it is unwise to try to separate them too much (Steenbergen,1996; 24, 30, 207-8).

*International issues – virtual water* The first section of this chapter demonstrated Yemen's dependence on imported grains. This is an economic "problemshed" solution for an environmental watershed problem. Allan (2000) contrasts the economic sense of importing virtual water in grains with the political nonsense of food self-sufficiency and points out that governments reconcile the two by ensuring the former remains a silent fact and the latter (only) audible rhetoric simultaneously. This self-deception is possible because the truth of ships, docks, mills, trucks, warehouses and thousands of people collecting sacks is the unsanctioned, and hence, silent partner. A few minutes of a key speech by one man which contains a few stirring words about the harnessing of water resources by our great farmers and engineers is sanctioned and hence audible:

> I have instructed the government to reconsider the recent increase in the price of diesel fuel so that we can encourage farmers to expand their planting of agricultural produce..we do not see anything to prevent us from success in attaining self-sufficiency from locally produced food in time...
> (President Ali Abdullah Saleh, 20th January 1996, ROY Radio, San'a)

The process of wheat and flour importation is also highly politicised. Grain and flour subsidies through exchange rate differentials on "essential imports" have previously selectively benefited those traders and politicians involved (Financial Times, 8/3/95). Irregularities have, at various times, reflected an attempt to drive people away from the cheaper government distribution points to the more expensive black market. Control of local distribution points, at least in Ta'iz, was transferred from opposition to ruling party in the run up to the 1997 elections when votes seemed to be won by the sackload (of flour provided).

## Conclusions: Actors, Allocations and Asymmetries

The history of water allocation between Ta'iz and Habir/Al Hayma has demonstrated a multi-layered political arena in which local and national issues become inextricably enmeshed. Despite the party-political exterior, local political reality is about individuals, their grouping and splitting over issues and the single constant of self-preservation and nest-feathering. Political networks are complex and clan related and dominated (but not exclusively so). Political parties are manipulated by local political actors to their own ends and vice versa and 'events and struggles at the local level can have a momentous impact on... the state' (Migdal, 1988; 36). Governors, ministers and even the President have become embroiled in the struggle. The issue of water reallocation has demonstrated that the political boundaries or affiliations and initiatives are:

1.     complex,
2.     are dominated by the ever-changing networks of relationships, and

3.  always have an historical context.

Borders are not watersheds. This is particularly apparent with the Ibb-Ta'iz border at Dhi Sufal. Although problemsheds are more appropriate than watersheds in the regional context of domestic, industrial and food needs, watersheds are still the locally immediate context, and typically reflect current upstream – downstream inequalities.

Shifting the potential source area for Ta'iz's water has mainly followed the existing power asymmetries, although this argument is somewhat circular in that being able to direct water allocation is perhaps the ultimate indicator of power in a water-scarce region. The legitimacy of water reallocation is the subject of the next section.

## Implications of Legal Pluralism for Water Allocations in Yemen

Trade in water is enabled on the basis of rules and regulations (Coase, 1992; 718). A complicating factor in Yemen, is the plurality of legal frameworks under which they operate which itself is identified as a weak state / strong society phenomenon (Migdal, 1988). The absence of a water law, the institutions to enact it and the ability to enforce it demand an alternative means of agreed allocation of Yemeni life's most vital commodity – water. This section suggests that in Yemen, where local forms of (customary) law and courts (via the shayx) are predominant, legal pluralism is particularly appropriate. This section seeks to summarise the different sources of water law in Yemen and briefly discuss the application and effectiveness of those laws regarding water allocation, the markets they regulate, and the rules by which they regulate them.

### Sources of Law

Four sources of law may be identified in Yemen; the Constitution, the Shari'a or Islamic law, Civil Articles and Customary Rules (Al-Eryani et al, 1995 and Haddash, 1998). The Shari'a is the ultimate source of law and the other sources are meant to be based on it. However, new legal problems arise with social and technological changes (Vincent, 1991; 200) which the days in which the Qur'aan was given neither foresaw nor legislated for. All four sources are therefore dependent in origin on its interpretation, which is a combination of qiya (deduction by analogy) and ijma' (consensus). The four main schools of Islamic law in Yemen (Hanafi, Maliki, Zaydi and Shafi'i) interpret Islamic law differently, which contributes to differences in the other three sources of law. Maktari (1971; 3) identifies the influences of these schools in Yemen at different times and in different places, the Shafi'i school being of particular significance in his study of water rights in Lahj.

The Yemeni constitution is based on the Shari'a and the civil law is its formulation. Civil law 'serves as a legal instrument to implement certain interpretations' which 'are disputed among the various Islamic schools' (ibid; 43). Similarly, customary water rights cannot be considered in isolation 'since customs have to adhere to Shari'a' (ibid.) and again, some differences in customary law occur as a result of the law originating from different Islamic schools. Customary law is portrayed as the local adaptation of the shari'a and civil law, moulded by the specific conditions of use (historical and geographical) and is defined as:

> the continued repetition of certain actions or practices by a collectivity in the conviction that they are legally binding. (ibid; 43)

Maktari identifies two types: 'adaa:

> The repetition of a thing [an action] invariably or mostly on the same pattern without reasoning. (ibid; 5, 6)

and 'urf (which is the usual term used in the literature and in the Ta'iz locality, whether this meaning is intended or not):

> that which human nature accepts by reasoning, and is acceptable to man's nature or habit. It is also authoritative, but it is readily understood.

Codification of customary law is rare, however the three hundred-year-old "Document of Seventy" cited in Al-Eryani et al (1995; 44), the law of Sultan Fadl from 1950 (Maktari, 1971; 69) and the five-hundred-year old customary laws of Al Geberti in Wadi Zabid (TESCO, 1971; 5) are examples of its codification.

*Legal Applications to Water Allocations*

A summary of the relationship of the various sources of law operating in Yemen to the shari'a with respect to some of the more prominent water rights as considered by those sources is given in Table 4.2 insofar as they can be tabulated. The table is intended to summarise Al-Eryani et al (1995; 41-77) and provides some of the basis of the plurality of legal stances in Yemen regarding water rights.

*Water ownership* Major differences between the legal systems occur with respect to water ownership. "There are no customary rules which explicitly define the legal status of water" but there is a distinct contradiction between the constitution and the shari'a and also between the constitution and the civil law (ibid; 46). The constitution uncompromisingly considers all water to be the property of the state (Haddash, 1998). The constitution's position appears untenable on this point. Although of "state interest", water in the view of the civil law and the shari'a (and therefore customary law) is *res nullius*, ("mubah", that is, "of nobody").

*Receptacles and wells* A key issue in water ownership is whether the water is in a receptacle. Civil law considers water not in a receptacle to be state property whilst that in a container to be the property of the owner of the container. Thus, once the water is contained it becomes private property and hence saleable. Although this contradicts the constitution, some legal apologists try to construe water "sales" as simply compensation for the cost of production. The second key issue stems from the question 'at what point can water be considered to be contained?' and specifically whether a well is a container. The Hanafi school and, for that matter the Law of Gravity, do not consider a well to be a receptacle, though, hydrogeologically, it may be considered a leaky one. That is, to be an effective well water must be able to leak into it, and if water can leak in it must be able to leak out. This is the fundamental difference between a well and a cistern. The rest of the Islamic schools and all the Yemeni sources of law do consider wells to be containers, and hence the water inside them to belong to the well owner. Hence, as a molecule of groundwater travels the 1mm across the seepage face of the well from the aquifer to the inside face of the well surface, its legal status changes from "common" to owned. Put another way, 'an aquifer is freely accessible...to anyone who owns the overlying land' (Haddash, 1998). This provides a huge legal incentive to drill wells.

*Water diversions* Haddash (1998) identifies a distinction between ancient water diversion rights and more recent rights of "benefaction" where water is declared as state property. The constitution regards water diversion rights as a concession (which should therefore be regulated) whilst other sources of law submit to local custom. Civil law permits water from any source to be appropriated on the basis of seniority of claim and absence of harm to existing users (ibid.). Regarding the linkage between water and land, one interpretation of the shari'a ties water to the land and not to the landowner, whilst the other separates them.

## Table 4.2 Relationships of legal systems to the shari'a

|  | Relationship to Shari'a | Codification |
|---|---|---|
| Constitution | Principles | Constitution |
| Civil Law | Modern Formulation & Interpretaton of Schools | 1399 Articles |
| Customary Law | Local / Regional Instrument of Implementation | Document of 70 |

Summary of Water Rights

|  | Includes | Constitution |
|---|---|---|
| Ownership - Recepticle | Legal Status of Water | State Property |
| Non-Recepticle | & ownership conditions | State Property |
| Water Diversion | Basis of Right | Concession to User |
| (surface & groundwater) | Transfer of Right |  |
| Water Use | Priority |  |
|  | Quantity |  |
|  | Place (Water Transfers) |  |
|  | Deficit |  |
| Water Administration | Allocation Systems |  |
|  | Operation & Maintenance |  |
|  | Quantity & Quality Protection |  |

Summary of Rights of Servitude

|  | Shari'a |
|---|---|
| Irrigation Rights | Privately Owned, water / land divorce possible |
| Drinking Rights | 1.Drinking, 2.Animals, 3.Domestic, 4.Irrigation |
| Water Way Right |  |
| Drainage Right |  |

*Source* : As marked on next page

## Table 4.2 Relationships of legal systems to the shari'a cont...

| Water Related Articles |
| --- |
| 38 |
| 3 |

| Civil Law | Shari'a | Custom |
| --- | --- | --- |
| Private Ownership | | |
| State Property | Common | |
| First come first served subject to custom | | |
| Water 'married' to land and transferred with it | water 'divorced' from land and can be sold or inherited separately | |
| 1.Drinking, 2.Animals, 3.Domestic, 4.Irrigation | | |
| Prior users excess is available to others (based on original use) | | |
| Water transfer banned if anothers' water right is harmed | | |
| | time shares proportional to land | |
| | time shares proportional to land | |
| Costs according to use share, protection from channel | | |
| Protection Zone Declared | Principles of Pollution Protection | 1. Document of 70 - protection zone = well depth. 2. Drilled 500m, Dug- no rules |

| Civil Law |
| --- |
| Privately Owned but 'married' to land |
| Water transport across land permitted for fair compensation |
| Various rules to protect from flood damage and from downstream pollution |

*Source*: After Al-Eryani et al, 1995; 41-77.

*Water use and administration* Common to all the sources of law is that for drinking purposes, all water is *res nullius* (common or mubah) providing the desperation of one party does not annul the rights of the other party (Al-Eryani et al; 58). However, this stance is very minimalistic, as Ward and Moench (unpubl.) point out. It also directly challenges the practice of selling drinking water. For instance, a prominent Yemeni water bottling company, when asked by supporters of the shari'a why it sold water, replied that they did not sell water, just the bottle containing it. An interpretation of the shari'a forbidding the sale of water could also be an obstacle to intersectoral water transfers.

The free nature of water does not apply to all uses, and water for irrigation is not mubah if the new user will harm the senior benefactor (Haddash, 1998). Under the shari'a surplus well water is free to be exploited on a seniority basis (Ward and Moench, unpubl.). The problem with this rule lies in determining what is a surplus. Surplus water could simply mean that water is present in a well irrespective of a long-term decline in water levels. If non-declining resources are to be determining the presence of surplus then the need for monitoring is implied. At different times (Caponera, 1973; 213) and in different places (TESCO, 1971; 5) the generally ubiquitous "'ala fil 'ala" (upstream first) rule for stream and flood flow has been reversed by local custom. Serjeant (1964; 55-56) also gives examples of local jurisdictions regarding the operation and maintenance of water channels based on customary law. Whatever system of stream and flood flow allocation operates, the supervision of the system seems to be usually by locally appointed figures who become more necessary and more obvious in times of shortage.

Regarding the protection area around a well, spring or flood course, the legal sources differ (Table 4.2). The oft-quoted 500m dug well and 1000m borehole spacing rule is a ministerial decree and not actually "law" (Ward and Moench, unpubl.). Customary rules include the Document of Seventy that cites a protection radius around hand-dug wells equalling their depth.

*Rights of servitude* Rights of servitude, of which the irrigation right is a type (Haddash, 1998), are defined as 'the right of a certain property to obtain some kind of service from another property' (Al-Eryani et al 1995; 53). The four rights concerning water and the respective position of the shari'a and the civil law are given in Table 4.2.

Five main outcomes of the rules described above are noted (developed from Ward and Moench unpubl.):

1.    Free access to any source for individuals and livestock.
2.    Upstream priority over spate flows.
3.    Landowners can drill anywhere on their land and abstract as much as they wish.
4.    Recognition of protection against "harm" by new abstractors over existing ones.
5.    Some form of compensation for water traversing another's land.

## *Local Reality and the Problem of Law Enforcement*

Local reality dawns the day a drilling rig turns up in the neighbour's field. Mechanisms for judicial conflict settlement are in place at national (Supreme Court), governorate (Appeal Courts) and local level (Primary Courts) as are arbitration systems, that is, civil (using the judiciary) and customary or tribal (using the local shayx, up to shayx al mashayix level). The prior well owner, correctly thinking there is a moratorium on drilling without permit, will typically approach the SURDU extensionists or the Agricultural Office. In order to get an engineer to come out and confirm that there is a well being drilled, there is no permit and the new well is too close to his own well, he has to pay. The engineer's report is issued and strangely finds all these facts to be the opposite, because the new well owner has since paid a visit to Ta'iz and "convinced" the civil servants otherwise. The senior well owner is not then likely to approach any government court because he knows that government law, or rather judges, will lean to the highest bidder and will take a long time over it (Vincent,1991). The disputing farmers may then seek a local shayx whom they both trust to give a fair jurisdiction. The jurisdiction may involve supplying the senior well owner with water if his well is affected or promising to cease pumping if a certain water level is reached. Although both parties must agree to abide by this jurisdiction whether this happens once pumping begins is another matter, but the impasse over drilling may have been overcome. Other would-be abstractors misuse their government contacts to simply conduct their drilling by military force with no attempt at a negotiated settlement. These accounts are intended as a "representative case history" pieced together from several cases observed at different stages during fieldwork to the north-west of Ta'iz some of which are reported in Handley (1996a).

Another notable instance of particular importance to Ta'iz is where the "junior" well driller is the government. This case history was discussed in detail above, but a legal aspect is worth mentioning here. The ability of local well owners to block government drilling even when supported by military intervention demonstrates that the more local (bottom-up) the process of arbitration (that is, local customary law, court location and arbitrator), the more likely an acceptable solution will be found and law enforcement will be possible. This observation is also borne out by the instances given for Lahj (Maktari, 1971; 131) and Wadi Dahr (Mundy, 1989; 109).

Similar accounts of bribing government officials, injustice in the courts and local resistance to externally imposed rules are observed in most other areas of life such as paying tax, building a house, finding a job etc. The failure of the public water supply utility, NWSA, to carry out its legal responsibility to collect revenue is not always an institutional failure but, as in the case of powerful non-payers is sometimes a problem of law enforcement. These problems underline the importance of the presence of an indigenous albeit rather *ad hoc* or customary legal system. This locally-relevant, cheaper, quicker and probably more just alternative makes a far greater contribution to conflict settlement than does the government system and demonstrates that legal pluralism can be advantageous.

## *Legal Framework Implications for Institutions, Markets and Water Resources*

Although the jurisdiction of the Yemeni central government legal framework in conflict resolution may end at the San'a ringroad (Dresch, 1989; 16), it could have a major role in water resources protection. The effectiveness of two types of institution with respect to the Yemeni central government legal framework has particular significance for water resource sustainability. One is the institution that creates the law and the other is the institution(s) that monitors and enacts it. Despite many calls for a water law to regulate exploitation (Haddash, 1998, Vincent 1991; 201) and the writing of several drafts (World Bank, 1993; 16), none has actually been forthcoming. The ineffective High Water Council and its Technical Secretariat have been superseded by the creation of NWRA which has been invested with responsibility for water resources strategy, planning and regulation (Ward, 1998). NWRA's ability to win the trust of local well owners and involve them as stakeholders in the policing of their own resource is likely to largely determine whether a passed water law can actually be implemented.

Where the national legal and institutional frameworks have failed to ensure allocatively efficient use of water the markets have fared better. Some question the legal validity of markets, or at least suggest they need to be legalised (Dellapenna, 1995; 154) whilst others see their existence *ipso facto* as sufficient validity (Ward; 1998). The latter view is perhaps more applicable in a weak state/ strong society such as Yemen where the legal validity of an existing Government or formal institution may be perceived as a largely irrelevant issue.

The sale of irrigation water at different prices for different crops may set a precedent for customary rules regarding water sales on which to base negotiations for water transfers (Ward and Moench, unpubl.). However, irregularities in operating procedures by markets beg the intervention of some form of regulation. The urban public is particularly in need of protection from over-pricing, poor water quality and public health threats from water supplied by tanker, water treatment companies and even by the public utility.

The exploitation of water resources by operators within water markets does not protect the resource. In addition, it neither provides water for the poor at the demand end nor protects livelihoods of those not directly involved in the transfers at the supply end. Access to water resources is controlled by a small, relatively wealthy portion of the population (Ward and Moench, unpubl.) and well owner's claims of exclusive rights over groundwater abstractions meets with fierce opposition by other farmers (NWRA, 1998a). The legal incapacity to protect the environment where government law does not hold sway and to provide equity form a major obstacle to the sustainability of any allocatively efficient measures which might be implemented in the Ta'iz area and is a type of second order scarcity (Ohlsson, 1999).

The relevance of the legal empowerment and validity of institutions and markets for the potential to establish water transfers, particularly intersectoral transfers, are likely to depend most on the predominant social setting. Some role for central government regulation is envisaged, particularly in the provision of equity in the urban setting. However, beyond the Ta'iz ringroad, where any water transfers are to be sourced, the local customary legal framework is predominant and government civil law enforcement is limited.

Yemen is well known for its legal pluralism. This pluralism extends as far as water law. The players in the water allocation game not only face the moving goal posts of development initiatives and technological change, but also different sets of rules under different referees. Some players use pluralism to advantage, and law aimed at providing equity and order may sometimes miss the mark in this plural environment. However, the fact that several water allocation games are being played at once means that there may at least be speedier and more just legal alternatives.

## Institutional Appropriateness: Matters of Function and Scale

The plurality of legal frameworks in Yemen is reflected in a similar plurality of institutional frameworks in which they operate. Several games of water allocation and several sets of players are involved. The problem of water scarcity reveals itself to the public in the form of inadequate provision of services. Although some productively efficient solutions can be applied, these immediately perceived problems are a subset of the wider issues of allocatively efficient solutions and, when food needs are included, of virtual water solutions. Both the immediate and wider problems and solutions have an institutional context. As evidenced in the first section of this chapter, the problems and solutions at all scales involve water transactions. Coase suggests there is 'little sense....to discuss the process of exchange without specifying the institutional setting within which trading takes place' (1992; 718). In this section, some of the institutions involved in water management in Ta'iz are described and their contribution to the problems and solutions discussed. In conclusion, the possible future role of institutions in the water sector is briefly considered in the context of current development initiatives.

### Introduction: Immediate Problems and Broader Issues

It is suggested that the immediate water related problems and issues facing the general public in Ta'iz centre around the provision of the following:

1.      urban domestic and industrial water supplies,
2.      rural domestic water supplies, and
3.      waste water disposal facilities.

In particular, these services need to be provided at cost recovery / economic prices, whilst at the same time safe-guarding public health, and promoting productive efficiency so as to protect the resource from over-abstraction and pollution.

In addition there is a problem of enabling steps towards intersectoral, allocatively efficient water use, through water transfer arrangements negotiated with sufficient support of the stakeholders to ensure conflict is minimised.

[Productive efficiency in the irrigation sector has not been included in the above list, partly because:

1.      it does not serve to protect the resource, and

2.      the institutions involved in irrigation are also involved in the problems mentioned above and are discussed in that context.

Also, the water aspect of securing an adequate food supply is not discussed in this section primarily because it is a political issue rather than an institutional one. The institutions needed to ensure food provision, via virtual water, are in place and functioning reasonably well.]

*The Institutions*

There are many institutions involved in the water sector in Yemen. However, in the context of the specific problems of water supply and disposal and allocative solutions, some institutions are much more relevant than others and greater emphasis is given to them here. The institutions which have overseen the emergence of the problems and which have attempted to address them are described below with respect to:

1.      their contribution to the problems,
2.      their contribution to the solutions,
3.      their adequacy as service providers / reallocative enablers,
4.      the roles of institutional plurality, and
5.      institutional adaptive capacity.

These issues are discussed in the conclusions. In order to avoid a narrowness of scope it is helpful to divide the institutions into Government / non-Government and Traditional / non-Traditional. The government / non-government distinction is made in preference to a formal / non-formal division since the origin of an institution is far more significant to the Yemeni, than its degree of formality. Institutional formality is a function of legal status which was discussed in the previous section.

*Government Institutions*

The relationships between central government ministries and their departments and the specific offices and projects found in Ta'iz and externally derived donor involvement are summarised in Figure 4.2 and their responsibilities in Table 4.3.

A chronological sequence of involvement of various institutions in aspects of water resources management in the region may be discerned, and is reflected in the order in which they are described below. Institutions with secondary involvement are only mentioned briefly below.

**Table 4.3  Responsibilities of government institutions involved in the Ta'iz water sector**

| Institution | Responsibility |
| --- | --- |
| **Rural and Urban Role** | |
| Governor's Office | Government of Ta'iz governorate, including civil law and order. |
| NWRA | Water resources policy and management. |
| TWSSP | Financially, technically and institutionally facilitate water transfer to Ta'iz. |
| **Rural Emphasis** | |
| Agricultural Office | Execution of agricultural law; dams and irrigation schemes; other agricultural fields. |
| AREA | Research and extension for agriculture. |
| CACB | Provision of development finance for farmers. |
| GAREWS | Rural water and electricity supply. |
| IDAS | Enhance self-help capacity of farmers. |
| Local Councils (LC's) | Promote and manage local, rural development including water supplies and irrigation. |
| LWCP | Implementation of water monitoring, irrigation, and forestry projects. |
| SURDU | Rural development including irrigation, extension, monitoring. |
| **Urban Emphasis** | |
| NWSA | Urban water and sewerage supply. |
| TSWSSSR | Help NWSA branches towards a more autonomous and commercial basis, and encourage PSP in urban water supply and sanitation provision. |

*Source*: Developed from Hansma and Hermans 1997,Table 5.1.

*LDA's − LCCD's − LC's*: The following comments are based on Joshi (1995). Development in rural areas and small towns has centred on Local Development Associations between 1973 and 1985, Local Councils for Co-operative Development from 1985 until the early 1990s and Local Councils currently. In North Yemen, over 200 LDA's were established by the early 80's and were two-thirds locally financed, one-third government financed. During the period 1973-1984, 4,244 water projects were constructed.

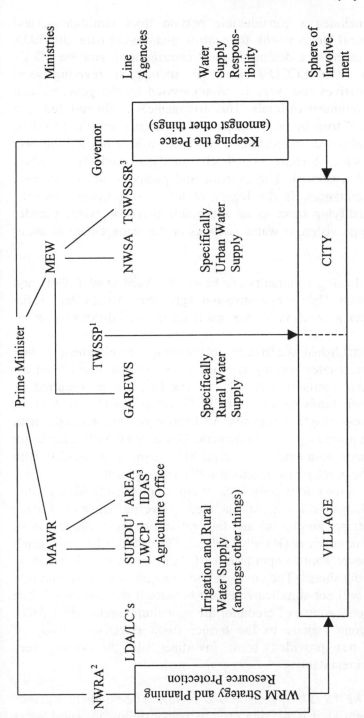

**Figure 4.2 Government institutions and donor projects involved in the Ta'iz water sector**

Local finance included a considerable portion from remittances, and attempts by central government to control and co-ordinate the LDA activities led to mistrust, a decline in local contributions, and the LDA's eventual demise. The LCCD's were an attempt at reviving local development initiatives and were to be supervised by the governor and other central government officials. This arrangement again resulted in a well earned lack of trust by the principals of the agents, and the LCCD's, faced with no real power to create or administer projects beyond the firm control of the state, became completely paralysed and its members disenchanted and frustrated. The mistrust and paralysis, not to mention drying up of remittances, is the legacy of the current Local Councils, effectively disqualifying them as an acceptable player in water transfer negotiations, the provision of water facilities or the management of water resources.

*CACB*:   The following comments are based on Ward et al (1998) and World Bank (1998a). The Co-operative and Agricultural Credit Bank is the third largest branch network in Yemen and is the main credit institution for agriculture.

It was established specifically to encourage development in that sector. With interest rates running typically above 30%pa, loans from the bank are particularly attractive at 7-9%pa. The LCCD's are required by law to deposit their funds with the CACB. However, the CACB does far better placing these "captive deposits" in interest bearing accounts with other banks, than providing loans to farmers. 43% of the CACB's funds are long term soft loans from external sources. Most loans from the CACB to farmers are for the purchase of irrigation wells or equipment.

Loan recovery rate is only 60%. Were it not subsidised, the bank would have to charge 59% interest. The bank is overstaffed (only 9 loans allocated per year per employee) and farmers accuse those employees of principal-agent irregularities (Handley,1996a). The result is that most small farmers would never want to approach the CACB for a loan. Outstanding debtors are the "big shots". The courts refuse to adjudicate when land title is collateral and will not adjudicate interest because it is "unislamic". The result is that the only source of credit for the agricultural sector, the CACB, provides the wrong signals to the farmer (with effective subsidy of irrigation) and has provided scant investment in the sector very inefficiently and inequitably.

*SURDU and GAREWS (LWCP, AREA and IDAS)*:   The Southern Uplands Rural Development Unit is passing over its responsibility for rural water supply to the General Authority of Rural Electricity and Water Supply. However, the latter organisation has a track record of not completing the village water supply projects it began. Problems of lack of maintenance of projects that were completed and of constructing "blueprint, top-down" schemes without reference to village needs are also common (Handley, 1996a; 26 and Handley, 1996b; Table 3).

The Land and Water Conservation Project, the Agricultural Research and Extension Authority and the Innovation Development in the Agricultural Sector project, together with SURDU are all involved in irrigation projects, and IDAS has also been involved in some minor rural water supplies (Hansma and Hermans,1997). These organisations all operate in the rural sector and there is considerable overlap and duplication of responsibilities between them. 6.4% of the rural population of Yemen are provided with public piped water supplies and only 3.8% in the Ta'iz governorate (Table 3.13) suggesting this key activity is very inadequately and inefficiently addressed.

SURDU and the Agricultural Office are accused by farmers of principal-agent irregularities when determining whether a farmer can drill a well near his neighbours. The recommended 500m spacing for dug wells and 1km for drilled wells appears to be flaunted or adhered to depending on who is the highest bidder.

*NWSA:* The National Water and Sanitation Authority was created in 1973 and is part of the Ministry of Electricity and Water. Some performance indicators are given in Table 4.4.

**Table 4.4  National and Ta'iz NWSA performance indicators, 1997**

| Performance Indicators | Nationally | Ta'iz Branch |
|---|---|---|
| Water Sold m$^3$/yr/employee | 12,000 | 6,500[a] |
| Connections / employee | 63 | 78 |
| Electricity as % of revenue | 21 | 60% |
| Wages as % of revenue | 43 | 60% |
| Outstanding Accounts | 14 months | ? > 6 months |
| Unaccounted-for-Water | 31% | > 44% |
| Losses $M/yr[b] | 1 | 0.34 |

[a] The Ta'iz branch has 462 employees of whom 46 are over the retirement age and 26 have died but whose families continue to receive salaries.

[b] No allowance for depreciation is included.     *Sources*: NWSA data and Handley, 1999b.

The table aptly demonstrates the consequences of treating water as public good. Branches have been obliged to charge a national tariff too low for cost recovery and branch revenues are sent to head office and re-disbursed to branches through non-transparent budgetary processes (Davies and Sahooly,1996). Branches are overstaffed, there are no job descriptions, standards or codes, and performance or merit play little part in rewards or promotion (ibid.).

The knock-on effects are inadequate pay, lack of job satisfaction, a poor working environment and a declining skills base so typical of weak institutions (Coopers & Lybrand, 1992). Many staff need a second job and there is widespread "rent-seeking" and increasing unaccounted-for-water. In addition, a fundamental change is needed in NWSA, if it is to become more consumer oriented and the ugly scenes at the NWSA offices of consumer blaming supplier for shortages and vice versa are to be less frequent. Although these problems permeate most of NWSA, they are particularly acute in the Ta'iz branch.

The provision of sewerage connections in Ta'iz has not kept pace with water connections (Figure 4.3), which in turn has not kept up with population growth (Figure 3.8). The declining water resource base has resulted in less water being pumped, and one advantage of falling production is that losses are reduced (Figure 4.4).

Contrary to Berkoff's recommendation to use rotational deliveries in water scarce situations (1994), reduced water supplies in Ta'iz have been observed to result in a vicious circle of increased corrosion and hence increased leakage. Water rationing causes pipes to be exposed to air and water alternately resulting in more rapid corrosion, low quality water produces more rapid corrosion of storage tanks and lower flows in the sewerage network results in higher concentrations of sewage which corrodes the sewer pipework.

The worst legacy of NWSA, in the context of potential water transfers, is that government institutions are perceived as having lied to the people of Al Hayma regarding the affect of abstraction for the city on the farmer's wells. They are also accused of having struck a dirty deal with the local shayx in order to drill the city wells. This reputation has spread throughout the Upper Wadi Rasyan catchment and beyond, and is immediately raised as an objection to any proposals for exploratory drilling for city supplies. Contrary to Dellapenna (1995; 155), state system sceptics are not only justified in doubting whether experts can acquire the information to arrive at the right conclusions, in the Ta'iz case they are not sceptical enough. Even if the right conclusions are reached, will the state fully inform the stakeholders of the conclusions?

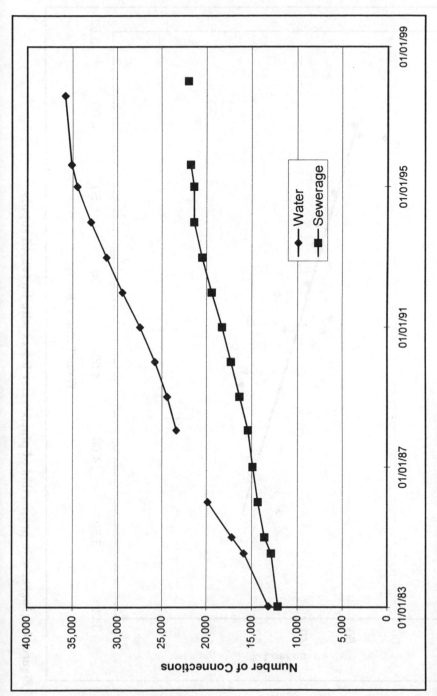

**Figure 4.3 NWSA water and sewerage connections**

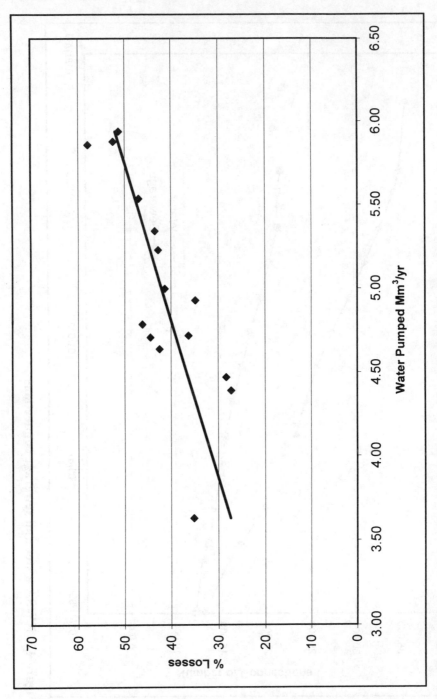

Figure 4.4 **Increase in proportion of piped water losses with increased supply**

*TSWSSSR:* Because of the institutional problems faced by NWSA, the Technical Secretariat for Water Supply and Sanitation Sector Reform has, in consultation with NWSA, embarked on a programme of sector reform. The programme has three main phases (Davies and Sahooly, 1996, Kalbermatten, 1998):

1.      The establishment of the branches on an commercially viable autonomous basis.
2.      The conversion of the branches to fully autonomous Regional Corporations.
3.      The encouragement of private sector participation in water supply and sanitation activities.

The first phase includes branch autonomy in establishing cost-covering tariffs and salary incentives and in hiring and firing. Community participation in the operation of the utility is also on the agenda, as well as reduction of unaccounted-for-water and the promotion of good customer relations and improvement in service quality. These aspects of the technical secretariat's reform agenda are at present lacking in the Ta'iz instance, and the branch has been singled out as too difficult to reform until after most other branches will have entered the programme (ibid.).

*The governor's office:* With the failure of the Al Hayma wellfield in the summer of 1995, the Governor of Ta'iz embarked on an "emergency" drilling programme within the city limits of Ta'iz. Although local businessmen were prepared to pay for the drilling, the Governor demanded payment by central government for the most expensive item, the submersible pumps, under threat of his resignation. The Governor was advised to monitor well drilling and yields. However this was not done, and the rushed completion of 18 wells resulted in an average well yield of only 3.5 lit/sec, which is still declining, (Dubay,1996) and derogation of existing wells. Incorrect connection of the pumps to the existing system and operational malpractice by NWSA has resulted in pumps running dry and overheating, wells pumping against each other and significant leakages. A more useful intervention by the Governor's office has been in negotiations with rural communities and shayxs over the water transfers to the city discussed above.

*HWC – NWRA:* Berkoff identified a typical historical development of water entities (1994):

1.      Water agencies are established to meet a specific need.
2.      Activities related to that use are delegated to autonomous entities reporting to the parent department.
3.      The agency covers all aspects of the water use and acts according to its own needs and biases.
4.      The result is competition and inefficiency.
5.      At this point an overarching intersectoral co-ordinating body representing the different ministries involved may be appointed.
6.      This body is delegated functions before it has the mandate, skills or resources to manage.

Yemen has been no exception to this trend. Table 4.3 and the above discussion indicate the considerable overlap between agencies in the rural sector and their inefficiencies. The recommendation to establish the overarching co-ordinating body occurred in 1980 (Berhardt et al) and the resulting High Water Council was established in 1981 (World Bank, 1993; 53). In fact it never sat (Ward, 1998) and point 6 above would have aptly described it. After the failure of the HWC, the National Water Resources Authority was established by Presidential Decree in 1996. This represents another attempt at creating the "overarching intersectoral co-ordinating body". However, whether direct accountability to the Prime Minister (also a feature of the HWC), the recruitment of "key" personnel from other agencies, and significant Dutch aid/UNDP inputs (as had the HWC), are respectively sufficient "mandate, skills or resources to manage" remains to be seen.

*Summary:*    Figure 4.2 and Table 4.3 indicate significant overlap of Government institutions in the irrigation and rural water supplies, and an emphasis towards irrigation which provides the wrong economic signals as far as allocative efficiency is concerned. In addition, productive efficiency is low in urban and rural water supply sectors and principal-agent irregularities reinforce mistrust in the Government institutions. There are "old agenda" O&M problems, particularly in rural schemes, as well as potentially productively efficient autonomy initiatives in the urban sector.

*Non-governmental Traditional Institutions*

Since traditions emerge and evolve, the distinction between traditional and non-traditional is loose and depends on the period to which the tradition dates back. The two particularly ancient institutions considered here, traditional support systems and the shayx, indicate both an existing and potential role in indigenous institutional water management.

*Traditional support systems:* Some of the traditional support systems of co-operation within and between communities identified by Joshi (1995) relate to water use. In particular, co-operation in irrigation is apparent. Indigenous mechanisms of water allocation relate not only to the distribution of collected rainfall, but also in the allocation of ghayl (stream-fed) irrigation observed throughout the Ta'iz region (Handley, 1996a; 23-4) and further afield (Varisco, 1983). Usually "invisible" to the observer, the "suruub al miiyaah" manager (water distributor) takes on a more prominent role as water becomes scarcer. Corporate response to water scarcity is an ancient feature of Yemeni rural life, a fact reflected more recently in the harnessing of remittances for water projects (for instance at Al Sina, World Bank, 1997) and also commercial ones (Handley, 1996a).

**Table 4.5  Traditional Yemeni support systems relating to water use**

| Support System | Description |
| --- | --- |
| Al-Ana or Al-shamla | Communal "voluntary" work. Penalties for those not volunteering. Well digging, bridges, dams, rebuilding after calamities. |
| Al-Muthaha | Mutual support among neighbouring farmers – irrigation equipment, labour etc |
| Al-Ta'awon fi Majal Al-ray | Co-operation in irrigation – one farmer is responsible for distributing water collected during rainfall. |

*Source*: After Joshi, 1995.

*The shayx:* Probably the key institution in water management (and most other affairs of any importance) in the rural area is the shayx. This style of informal self-government, although not as tribal as in the North of Yemen, typically runs in families until someone performs particularly poorly. The shayx is usually a senior member elected by local families on the basis of his knowledge of customary law and shari'a and, ideally, for his maturity and impartiality. The literature and fieldwork are replete with accounts of principal-agent and preference aggregation abuses (Johnson, 1996 terminology) by some shayx (see political section) although others may better represent their principals (Handley, 1996a; 24, 26). In either case, and at all locations visited during fieldwork, the shayx's word was final, a fact crucial to the success of the more recent non-traditional initiatives described below. The shayx is the management domain (Manzungu, 1999; 164) and cannot be ignored or bypassed in any stage of water management initiatives without the likelihood of jeopardising the initiative. It is vital to recognise their power, the limits to it, to inform them, to win their confidence and to work through them. They are the local political reality. Even with the recent trends of democratisation in Yemen, the inability of central government to implement law and order and collect taxes points to a continuing role for the shayxs in local government. Many of the more prominent "shuyuux al-mashaayix" (leaders of shayxs) also get involved in national government (Joshi, 1995).

*Summary:* The relevance and role of these two non-Government, traditional institutions stand in direct contrast to that of the Government institutions. Although there is a tradition of co-operation this has been limited to irrigation and rural water supply. The shayx system is also open to principal-agent abuses, but can and must be worked through, rather than around.

## Non-governmental, Non-traditional Institutions

Recent years have seen the introduction to public water supply of two non-governmental, non-traditional institutions:

*TWSSP:* The work of the World Bank in the Ta'iz region has demonstrated that there is a tradition of self-help in local, indigenous rural development in general, and in water provision for agricultural and domestic purposes in particular (World Bank, 1998a; Annex 5). The relatively recent establishment of indigenous co-operatives in the fields of agriculture and irrigation but also other purposes (Handley, 1996b, Ward, 1998), indicates the continuance of that tradition, and provides a potential basis for future initiatives.

The Ta'iz Water Supply and Sanitation Project has been established by the World Bank in 1997 to help facilitate relief of the failing city's water supply. It has considered the co-operatives or such similar non-traditional institution as potentially capable of negotiating rural-urban water transfers and distributing compensation equitably (World Bank, 1998b). Apart from supply side aspects and incorporation of public utility reforms in co-ordination with the GTZ sponsored TSWSSSR, the project also seeks to promote stakeholder participation in rural-urban water transfers through the establishment of committees / water user associations perhaps formed from existing co-operatives. NWRA similarly seeks to promote the involvement of stakeholder groups in building consensus over water resource management strategies in the area (NWRA, 1998b).

*Private sector participation (PSP):* The major area of indigenous initiatives in water supply has been in the private sector. The potential for private sector involvement in urban water supply and sanitation is currently being explored to varying degrees in many parts of the world (Brook-Cowen, 1997). Partly encouraged by the reality of the water markets already operating, both the TSWSSSR and the TWSSP have recognised the potential of PSP in Yemen and Ta'iz specifically. As indicated in chapter three, markets have responded quickly to demand and provide a much needed service. The markets have adapted, and function more efficiently than the public sector. However, there are limitations to the extent to which the private sector can take on large scale water provision. Apart from relatively small operations by individuals and small companies supplying tanker and treated water, the larger companies are reticent regarding involvement in the water sector. The two largest industrial groups in Ta'iz have contemplated involvement in bulk water supply to the city, but, apart from needing technical assistance to do so, they also have the following reservations:

1. The shortage of water resources (how can water resources be managed if there are none to manage?).
2. The water quality problems (the scale of investment in treatment makes it unattractive).
3. Current perceived management and operation problems in NWSA.
4. The run down state of the current facilities.
5. Lack of clarity regarding how company management, operation and objectives would be able to integrate with or adapt to NWSA branch management operation and objectives.
6. Fear of public opinion and reaction if any problems occur.
7. Mistrust regarding whether the government would truly give the private sector a free hand.
8. No guarantee that they will be able to recover revenues.

Within the scope of PSP being discussed (ibid.), both groups envisaged their potential involvement being limited to service contracts, though the possibility of a concession for specific areas of Ta'iz was contemplated.

Options requiring any greater investment or involving more of the city were considered too risky.

In addition to PSP in the urban sector, the largest company in Ta'iz has provided water and electricity supplies to villages as compensation for environmental damage and livelihood deterioration, as a promotional tool, or simply as a form of "zakaat" to needy communities.

Private businessmen working in the Gulf and Saudi Arabia have also made a considerable contribution towards providing water supplies in some rural home areas, on both commercial and philanthropic bases. Despite these private initiatives in rural and urban areas, if the utilities were privatised, questions of equity provision and the importance they would give to protecting the environment are still relevant (Hildyard, 1998; 18 and Allan, 2000; chapter 3).

*Summary:* Two major non-governmental, non-traditional institutions may be identified as currently involved in water supply in the Ta'iz area. The World Bank initiative has established the only institution directed specifically at the problem of reallocating water from rural to urban areas through stakeholder participation possibly via co-operatives or newly formed water user associations. Although the private sector has made some contribution to improving water supplies in rural areas, there is more reticence to get involved in large-scale urban supplies.

## Discussion – the Combination of Scale and Function

*Scale*  It is suggested that the appropriateness of institutional arrangements for the efficient provision of water needs might best be examined from the viewpoint of the scale and function of those arrangements. Regarding function, water provision may be divided into productively and allocatively efficient provision, with productive aspects subdivided according to sector. Regarding scale, institutions may be divided into village (or even Wadi), city (or regional) and national. In the application of institutional theories to Yemeni water problems, there is a distinction on the basis of the scale from local to national institutions. These distinctions are related to the urban / rural location in which the institutions seek to address the water management problems described above.

The national (government) institution is perceived, not undeservedly, at the local level as untrustworthy and as only being interested in "stealing our water – and then wasting it". Those who think that centralised institutions have good potential for co-ordination and integration (Guggenheim, 1991) obviously have not worked with NWSA.

Similarly those who prefer the state systems, because they are supposed to deal with problems more rationally before they become crises than can the private sector (Dellapenna, 1995; 155), obviously have not lived through the Ta'iz water crisis. The "participation pessimists" correctly favour legal reforms to improve the efficiency of formal institutions. However, if they discount the non-formal, local, institutions when it comes to the most important issue of water transfers, they will become non-participation pessimists. Part of the solution, it is considered, is the scaling down or decentralisation of the national, government institution towards something more amenable to, and perhaps even representative of, local interests.

Locally, one of the problems of negotiating water transfers, or for that matter of providing rural water supplies, is finding a "representative" body with which to negotiate. Relative to the urban situation, rural societal structure and responsibility are more clearly demarcated. Even so, the "participation optimists", who believe that non-formal institutions can transmit community demands, may prove to be participation over-optimists. As mentioned, both principal-agent and preference aggregation problems can occur when negotiating with some shayxs. Also, on a wellfield scale, the size of local institution (shayx or whatever) is often too small, as in the Hayma-Habir example, to represent those stakeholders who would be affected by water transfers. As a result, the "mish-mash" of local web-like institutions and political borders have to be bridged in the negotiations, with considerable addition to transaction costs and the risk of not arriving at any settlement with all interest groups. Part of the solution is the scaling up of local institutions into a representative body with which to negotiate.

Both activities are needed, the scaling down or decentralisation of the government institution and the scaling up of the local micro-institutions into a representative body at a more "catchment-based" scale.

*Function* Ta'iz people relate differently to water; as users for drinking, domestic use, industrial use agricultural use and as traders (or potential traders) in water, and the institutions centre around these water functions. The existing government institutions have been inadequate in their provision of drinking water and sanitation due to a combination of institutional overlap (in rural areas), inefficiency and principal-agent irregularities. In providing for irrigators, they have transmitted the wrong economic signals in terms of allocative efficiency. The industrial sector has essentially met its own water needs without recourse to outside institutions, and small scale trade in water has also been self-regulating.

To enable larger scale water transfers from irrigation to allocatively more efficient uses Hansma and Hermans (1997) suggest the manipulation of existing institutional webs which relate to the water functions. The World Bank experiment with water user associations in Miqbaba and Al Hayma is an alternative attempt at the creation of new institutions, though based on traditions of self-help and co-operation. Both these models are forms of decentralisation away from Government institutional intervention. Regarding equity provision, the non-Government institutions, although still prone to principal-agent irregularities, are much more relevant, and trusted than government ones at least at the village scale.

Some suggest environmental protection is a central government function, however, self-limitation of abstraction and pollution from below is likely to be much more effective than authoritarianism from above in a weak state/strong society. Although 'urf prevails over shari'a, the concept of ijmaa' and local consensus, rather than the imposition of non-indigenous rules, are much closer to the cultural heart and is a more sustainable basis for decision-making. A key measure is therefore the involvement of the informal local stakeholders via indigenous institutions.

A more realistic and relevant function for central government is the provision of the right macro-economic signals instead of the wrong ones (diesel and wheat price, subsidised loans, agency technical assistance etc.) to promote allocative efficiency. Also, determination of safe yields and pollution limits, and instruction on how to monitor and interpret the observations may still best be provided by central government function via a regional agency.

## Conclusion – the Potential for Local Water Management

The respective titles of UNDP and World Bank projects in Yemen of 'Strengthening of Water Resources Management Capabilities' and 'Decentralised Management' reflects recent thinking in development involvement in the East. In both cases, institutional appropriateness is perceived as the target. If aid is not to result in even less being achieved and even more skimming off by bureaucrats, then the recipient institutions will need to be locally accountable. The TSWSSSR programme, aimed at decentralising the provision of water and sanitation services, at least in the urban area, directly addresses the need to make more efficient use of the water. The TWSSP initiative seeks to promote the equitable representation of the supply area communities in water transfer negotiations and attempts to bridge the urban-rural (and government agency-local community) gap between consumer and source areas. Both these projects seek local stakeholder participation in the entities they aim to reform / create.

*Potential for collective action* By 1995 the historical legacy of Al Hayma was a rightly placed mistrust of central government (via its agencies) which had spread far beyond the Upper Rasyan area. From 1995 the TWSSP initiative has experimented with promoting grass-roots stakeholder participation in water transfers.

In Steenbergen terminology the TWSSP is acting as a facilitator (and financier) of the second order transaction costs of institutional change (1996; 198). Those costs have been huge and time consuming (Ostrom, 1999; 201) and may yet prove too expensive to result in successful formation of institutions. There may have been a critical moment (contrary to Steenbergen, 1996; 202) in 1976, when the institutions needed to be in place to negotiate, or refuse, the planned end of irrigated agriculture (Leggette et al, 1977). Although critical moments can often only be detected with hindsight, near total depletion has resulted in relative water scarcity for most, bordering on absolute water scarcity for some.

Despite the comments of those who think that scarcity or resource depletion should galvanize people into collective action (Mahdi, 1986; 193, Thompson, 1988; 67), at least in (new) water acquisition (Uphoff et al, 1990; 30), cooperation has not been automatic (Lam, 1994; 282) in Al Hayma, or Al Malika (Table 4.3), or many other Yemeni water depletion situations. In the irrigation literature "new" water may be available through more efficient operation of irrigation schemes, but not where there are limited and declining groundwater resources and inefficiencies are already recouped by recirculation.

Why doesn't scarcity result in collective action in Yemen in general and in Al-Hayma / Ta'iz in particular? In the TWSSP experiment, many of the prerequisites for institutional change and operation (ibid, 34-37, 47, 304-6, Mahdi,1986; 192, Ostrom,1986; 607-9, 618) were pursued and constraints avoided (Mollinga, 1998; 246 et seq). It is difficult to imagine more investment in second order transaction costs in the pursuit of so little water. Perhaps never in the history of development has so much been spent by so many to secure so little.

Despite this, two crucial elements have been missing, it is suggested. The first relates to the purpose of collective action in terms of supplying water to Ta'iz. This activity is essentially perceived as an external imposition of insufficient economic interest to the source area inhabitants. The second is the basic lack of trust between potential co-operators. At the best of times mistrust permeates Yemeni society, and is part of the underlying belief systems discussed in the next section, mistrust has been reinforced very effectively by the Al Hayma experience.

For equitable water resources management, the development community will have to ensure that decision-making roles by stakeholders reflect the level of the latter's stake in water rather than simply the current power asymmetries. For instance, Ward (1998) identifies drinkers with a basic need stake, users with livelihood stakes and owners with rights at stake. These stakes reflect the different roles and levels of importance water holds in their lives. If local communities are to heed calls for water resource protection from derogation and pollution they, according to their stake, and nobody else, need to own and police it. Establishing stake-weighted institutions endowed with control and responsibility will be a severe test.

Many writers note the importance of "policy by process" as opposed to 'policy by prescription' in water management (Mackintosh, 1992). The Ta'iz case suggests the policy (if there has been any policy at all) has been 'get water to the city by fair means or foul'. The means has contributed to the mistrust. Although this 'policy by default' approximates more closely to process than prescription, process has meant opportunism (Mollinga, 1998; 239) in which the actors have been few, consultation negligible, and contingency the order of the day (ibid; 242, 263).

However, the irrigation literature is mono-sectoral and hence limited to the productive efficiency approach. Solutions to the Ta'iz water shortage lie in allocative efficiency and have to address much wider issues. The present relationship between local and government institution (upper part of Table 4.6), reflected in, is contrasted with a proposed model (lower part) in which most water management functions are managed locally (arrows indicate a cause-effect relationship).

At the national level, the task of providing the right macro-economic signals to promote allocatively efficient water transfers remains the key central government role, whether or not there is political will to adopt it. The government institution with the closest mandate to identifying and effecting allocatively efficient policy is NWRA. The non-transparent performance of government institutions in the export of water from rural areas to Taiz in the past has placed the fledgling NWRA under the communities' microscope.

## Table 4.5 Present and proposed institutional functions

### Present Situation

Scale

| | Local Institution | | Government Institution |
|---|---|---|---|
| **F** | 'take' mentality | ========➔ | inefficient water provision |
| **U** | | | |
| **N** | conflict over | ========➔ | unacceptable reallocation |
| **C** | 'stolen water' | | |
| **T** | | | |
| **I** | allocative inefficiency | ⬅======== | wrong economic signals |
| **O** | | | |
| **N** | resentment of Impact | ========➔ | oversight of nvironmental |
| | and blame of government | | disaster |

### Proposed Institutional Model

Scale

| | Local Institution | | Government Institution |
|---|---|---|---|
| **F** | | | |
| **U** | Efficient Local Provision | ⬅========= | technical help |
| **N** | | | |
| **C** | Acceptable Reallocation | ⬅========= | right economic |
| **T** | | | signals |
| **I** | | | |
| **O** | Environmental Policing | ⬅========= | policy guidelines |
| **N** | | | |

### The Contribution of Knowledge and Belief Systems to the Degree Discourse on the Allocation of Water is Sanctioned, and some Implications for Development

Whether politicians really have a free hand in policy determination has been questioned (Foucault, 1971, Mollinga, 1998; 20). An alternative model in which interacting and competing interests of the powerful define and legitimise acceptable discourse has been proposed (Tripp, 1996 in Allan, 1999a). Within these boundaries, discourse is termed "sanctioned" or permitted, beyond which lies the wilderness of "unsanctioned" discourse. The boundaries are blurred, vary for different communities and may change with time given an appropriate crisis or "window of opportunity". This section seeks to demonstrate that belief systems and knowledge define those boundaries, directly influencing the degree of sanction discourse on the allocation of water receives, and hence directly affecting the extent to which allocative measures can or cannot be introduced.

Beliefs about some issues that relate to water allocation commonly held in Yemen are briefly considered together with beliefs regarding related issues of the state, equity and the environment. The extent to which beliefs and knowledge can influence water policy is discussed, and from this vantage-point, various water-allocative issues in development are examined and the influence of beliefs and knowledge on the degree to which discourse on them is sanctioned is considered.

*Belief Systems: Facts vs. Truths*

The importance of belief systems in determining what is politically feasible in water allocation has been argued by Allan (2000; chapter 4). Before considering the prevailing belief systems in Yemen regarding water, and in order to avoid confusion over terminology it is important to make some distinctions. There seems to be a confusion in the literature (Atkinson, 1991; 53) between facts and truths. It is not surprising that this confusion may also be found amongst Arabic speakers, where "haqaa'iq" is commonly used to cover both terms. To improve the objectivity of the following analysis and discussion it would be appropriate to define these terms.

*Fact:* an observable (often measurable) phenomenon that could be verified by witnesses e.g. it rained here today or water flows downhill.

*Truth:* an opinion adhered to. For convenience we will subdivide truths into true truths and false truths:

*True truths:* opinions based on a reasonable interpretation of the facts e.g. water has value (an interpretation of the fact that tanker drivers pay x money for water at the well), or the national food deficit is made up by the import of virtual water (an interpretation of national food statistics).

*False truths:* opinions not based on an interpretation of the facts or based on an unreasonable interpretation of the facts e.g. water out of the tap should be free. (This interpretation is based on a combination of an unreasonable interpretation of the fact that rainfall is free and/or failing to interpret the fact that costs are involved in the installation, operation and maintenance of infrastructure needed to get rainfall fit for consumption from the tap.)

By definition false facts are termed lies, and when interpreted may yield false truths. The difference between true truths and false truths is, of course, gradational. The truthfulness of a truth is also confused when truthfulness is based on the strength of adherence to the truth, rather than reasonableness of the interpretation or the extent to which facts support the interpretation. The problem for water managers is that most of the truths they deal with lie towards the middle of the spectrum where the interpretations become more tenuous, mainly due to the absence of complete scientific fact and unknowns in the method of interpretation (Thompson, 1995; 25). The resulting uncertainty provides scope for the certitudes (ibid; 27, 28). There are also some instances of unreasonable interpretation of the facts, reasonable interpretations of false facts etc. Despite the problems of interpreting information, it is hoped that the distinctions between facts, true truths and false truths will aid the debate.

There are also types of truth with origins not related to the interpretation of observations. These include religious truths (derived from holy writings or revered interpretations thereof) and social norms and traditions. It is also useful to distinguish conscious truths, which are verbalised, and unconscious truths and partly conscious motivations (Giddens, 1984; chapter 2), which are assumed, and can often only be detected by outsiders. These latter forms are often the most powerful because they are not perceived. They can therefore only be challenged with difficulty and tend to be excluded from the sanctioned discourse. They are often ubiquitous, invisible and have been in place for as long as people can remember.

An issue distinct from categories of "truths" is the impact of truths on human behaviour. In some instances behaviour may bear no resemblance to the truths expressed or, put another way, be termed "voting with one's feet", or hypocrisy. For example, the motto "God, the state, the revolution" versus the reality "me, my family, my tribe", or depending on grain imports whilst emphasising food self-sufficiency.

## Some Water-Related Truths

Starting with religious truth, curiously omitting air, it is stated that 'people are partners in three (things): water, fire and grass' (translation, Haddash; 1998). The interpretation of this verse has formed a firm "belief status" basis for establishing the unreasonable perception of water as a free public good by some (ibid.). This position starkly contrasts the emerging view (at least amongst intellectuals) of water as an economic good.

The absence of discourse over virtual water's national security role is maintained by several commonly held truths including national food security based on indigenous agricultural output, the social elevation of meat-eating and the necessity of procreation for the provision of labour and income now and in one's old age. Respectively, these truths obscure the facts, and increase the demand for agricultural and domestic water. Also, the assumption that agriculture has prior right to water over industry is fostered by the unconscious truth of having one's heart in the soil and the village (Dresch, 1989; 133, 306-7).

There are many traditional belief systems regarding water which apply particularly to village life that caution the romanticism surrounding indigenous knowledge, which currently emanates from the belief systems of some branches of Western academia. Drinking too much water is generally believed to be harmful, particularly for pregnant women and for children with diarrhoea, and skin with boils, sores or measles is not washed (Ansell, 1980). The belief that men should get priority access to clean drinking water over women, and that women should collect water when there is no motorised transport and men when there is (Mclagan, 1994) raises the question of whether male politicians and policy makers hold a smaller stake in water issues than women.

Related to the need for clean water for religious ceremonial ablutions in Islam, there is widespread reticence in the Middle East regarding the use of waste water that has been recycled in a sewage treatment plant for domestic purposes. Other than highly expensive desalinisation of seawater, the recycling of waste water would provide the greatest potential future source of water for Ta'iz (Handley, 1999b).

Although the domestic use of waste water is a necessity for many water and financially wealthier Muslims, this option has been rejected by Ta'iz leadership on religious grounds, a clear example of belief systems determining water policy.

*Views of the state* "The state is not me, nor my family nor my tribe and therefore is to be avoided or manipulated for the purpose of society, my society". The belief that corruption and bribery, taught from childhood by the ubiquitous cheating in exams, are a necessary part of life particularly when dealing with government appears to be one of the "unconscious truths" of Yemen. Some even find religious justification for illegal water and electricity connections. Coupled with a poor track record of government handling of local water issues, it is not surprising that surveys of public opinion find a poor estimation of government (Hansma and Hermans, 1997, Handley, 1996a). Migdal (1988; 40) states that government earns the right to rule by providing services / meeting needs. In Yemen there is an expectancy that the government should provide infrastructure (Lackner, in World Bank, 1998b; Annex 8) and piped water supplies in particular, at no expense to the user. Dresch links this to the historical pattern of northern tribal extraction of wealth from the south, commenting 'the more one takes, indeed, the more one expects' (1989; 373). The "take mentality" has been the experience of many who work in development, leading some to conclude that Yemen is not ready for that development (Chaudhry, pers comm.). Although the expectancy for government to provide is prevalent, contrary to Migdal (1988; 30, 40) that provision does not seem to impart the right to rule. This phenomenon appears to be closer to the weak state that expects little of its citizens (Myrdal, 1968).

*Views of equity* Whether equity in water allocation and provision is everyone's goal must not be assumed (Mahdi, 1986; 194):

> And we raise some of them above others in ranks

and:

> with respect to sustenance, Allah has favoured some over other

(Al-Zukhruf 43:32 and Al-Nahl 16:71, respectively, Lichtenthaler, 1999, 19-20). The widely accepted interpretation of this verse (ibid.) amounts to a passive acceptance, even expectation, of divinely appointed poverty, or the divine lack of right of commoners and opens the way to the institutionalisation of inequality (Mollinga, 1998; 238, Thompson, 1995; 31).

A farmer drilled a well right over a spring source and then sold the water from a pipe to the same downstream users who used to get it free by right from the stream. The drilling took place by military force because the farmer had good contacts in the army. This is one of many instances of inequitable changes in water allocation in the Ta'iz area in which oppressed and oppressor accept the new status quo. Although conflict over water resources and allocative measures is extremely common in Yemen (e.g. Al Wahdawi, 17/3/98), once resource capture has occurred the injured parties seem all too readily to accept the new status quo. Because of gravity this tends to be the downstreamers losing out to the upstreamers (Varisco, 1983).

*Views of environment* Despite the comments of some (Hansma and Hermans, 1997; 35), there is a considerable awareness of many "lay" people regarding environmental issues. Low quality water and the pollution that causes it is a sore point amongst those who have to use it, although they often feel powerless to take corrective action. Similar awareness is evident amongst those faced with declining water levels, most of whom show a better grasp of hydrogeological principles than their Western counterparts, although there are a few "limitless underground lake" believers and awareness of how much water is lost in the reticulation system is low. Even urban domestic water consumers know when tap water is of insufficient quality to make a good cup of tea. Like their rural counterparts they are very particular about which water source is used for which purpose (Ansell, 1980) and have at least some awareness of declining water availability and increased pollution. Because these impacts can be seen, smelt and tasted, the layman is not totally disqualified from analysing the situation and commenting on it (Hajer, 1995; 10). However, the cause of water shortages is mostly blamed on declining rainfall (Handley, 1996a) rather than on human activity. Although shortages are felt more acutely in dryer years no long-term climatic trend is apparent over the monitored period, although the increased abstraction pattern is very apparent. At least declining rainfall is politically easier to accept than increased abstraction, since one has less responsibility for vengeance against the perceived offender (God) than if one's neighbour were the cause (Allan, 2000; chapter 4).

There is wide adherence to the belief that the construction of dams is the solution to the country's water resources problems (Handley, 1996b) a concept which even gains Presidential voice (Yemen Times, 11/3/96).

This belief may be related to images of Yemen's former greatness during the time of the original Marib dam which even has Qur'anic support. However, there is a tendency for dams to favour upstreamers over downstreamers, silt up, and encourage malaria and bilharzia. Small, already silted-up "dams" distributed equitably down the valley, where crops could be grown on the silt and the water does not remain on the surface long enough to breed the larvae and snails, would be much more effective. The Yemenis certainly believe in these structures – they are called terraces.

## Types of Knowledge and their Roles

Reasons why the foreign expert's advice may be ignored (Morton, 1994; 45) are part of a wider debate concerning knowledge which might best be introduced by an example. At an IDAS workshop, in Ta'iz in December 1995, a group of local experts and stakeholders were asked about the major water-related problems facing agriculture and the possible solutions. They had to first nominate the problems and then cast two votes for the most serious in their opinion, all by secret ballot. Because some of the wider causes (diesel prices, lack of drilling regulation, population growth) and initial steps to solution (monitoring, policing) were not included in the first ballot, a foreign development worker nominated them. However, they still attracted no votes, which all went to water use efficiency measures. This confirms the observation of Allan (2000; chapter 4) that productive efficiency is politically the cheapest form of discourse. The reason it is politically cheap is that it does not challenge the old knowledge or belief systems. The new knowledge, aspects of which were put forward by the development worker, is politically too expensive to consider and remained firmly unsanctioned at the workshop. Although the new knowledge can become "mutual" in that foreign and local expert can share it, (Giddens, 1984) that change has not yet occurred significantly in Yemen where belief systems prevail. A return to the post-colonial "developer knows best" is not being proposed here, but rather a hybridity of the best of new and old knowledge (Bryant, 1998; 14). A further complication arises because of the belief systems concerning knowledge. The accumulation of knowledge through monitoring does not seem important and knowledge gained is guarded closely. Attempts to obtain information, either primary or secondary, are treated with deep suspicion, particularly when Yemenis are requesting it. This does not bode well for development goals of institutional capacity building.

Since power and knowledge are synonymous (Atkinson, 1991; 61) an issue of equity aspects of knowledge is raised. Like most of the Middle East, the virtual water issue is publicly maintained as a Yemeni blind spot (Allan, 1997, Golman, 1997) and the topic of sustainability remains the enclave of the enlightened rich (Chambers, 1986; 10) in San'a, if it is anyone's. The absence of action over the polluted Wadis of Hidran and Rasyan reflects the absence of any environmental discourse to which their representatives could contribute. Adequate policy-changing discourse does not exist because the problem is in an area of new knowledge, it is unsanctioned and it is not perceived by those involved in discourse. Within the "risk society" terminology of Beck (1999), the risk for Ta'iz is that there is insufficient water to keep everyone fed, watered, clean and in jobs in the current status quo. The earlier chapters have demonstrated the "scientific" factuality of this "risk statement" (ibid; 76), however the political value of the statement lacks an adequately "new knowledge-informed" audience to facilitate the risk becoming a threat. No policy change will occur because the discoursers do not feel the risks.

*Implications for Development*

This chapter has proposed that the water shortage of Ta'iz can be explained by successive layers of causation physical, economic, institutional, legal, and political and this section proposes that belief systems directly influence them. Into this context the developer enters (Manzungu, 1999; 159) with her "new knowledge". The development of development in the post-colonial era has followed a sequence roughly similar to these layers. Morton (1994) and Jewitt (1994) note a shift in emphasis from:

1.      industrialisation / modernisation, to
2.      agriculture / integrated rural development, then
3.      structural adjustment of macro-economics, to
4.      institutional/human capabilities, and now, in the irrigation literature,
5.      policy reform (Manzungu, 1999; 10).

Although the environment is missing from this list, Serageldin (1994) points out a shift in emphasis in the water sector from the old agenda of providing household services to the new agenda of environmentally sustainable development, but wonders if the old agenda is still on the table.

The history of development in Ta'iz can be related to Morton's and Serageldin's models but is less complete. The provision of the Kennedy scheme in the mid 1960's and emphases on agriculture, evidenced by the plethora of donor supported agricultural institutions in the city, are old agenda. More recent structural adjustment of macroeconomics is seen in the activity of the World Bank at national level, and the latest development of institution/human capabilities is an emphasis of UNDP, World Bank and GTZ initiatives at national and regional level. One might ask, however, what happened to industrialisation, why is agriculture still on the agenda and why hasn't the new agenda of environmental sustainability appeared on the table? The answers, it is suggested, relate to the extent to which discourse about these issues is sanctioned. These questions and related issues are discussed below in an approximate order of decreasing sanction, beginning with the sanctioned and ending with issues unsanctioned even in Western discourse on development.

*Irrigation, diesel prices and water law* Irrigation has been the direct recipient of development aid in the past, via import subsidies, cheap loans, fruit import bans and diesel subsidies and indirectly through the government agencies in the irrigation sector. It also continues to be supported, at least locally by the World Bank negotiated compensation package in Habir, which includes wells and dams for irrigation (World Bank, 1998b). In fact, Yemen's proportion of total World Bank lending for irrigation is second only to Egypt's (Berkoff, 1994; 5). Yemen is much more likely than Egypt to use up the loans, and the groundwater the loans enable to be mined, growing amphetamines. The economic nonsense of irrigation and the extent to which subsidised diesel prices give the wrong economic signals to irrigators have already been demonstrated. Government support of this status quo is reinforced by the absence of a water law, facilitating enhanced nest-feathering for the land-owning ruling elite in accordance with the prevailing belief systems regarding equity.

*Water supply and sanitation projects* Even though such projects are "old agenda" the development agencies seem to commit themselves to them easily enough in Ta'iz, as well as in many other towns and cities in Yemen (Gitec Dorsch, 1989, SAWAS, 1997, Dar Al Handasah, 1997). The predetermining allocative logic (Falkenmark and Lundqvist, 1995; 214) of these schemes has certainly led to inequitable access. Although the trend towards commercialisation and privatisation in the name of institutional development may lead to improved operational efficiencies (Davies and Sahooly, 1996), it remains to be seen whether equity is addressed (Allan,

2000; chapter 3). Discourse regarding whether the provision of urban services creates jobs and attracts more people from the countryside, or put another way, urbanisation breeds urbanisation, seems to fall in the unsanctioned category. The high cost of sanitation provision (Serageldin, 1994) often ensures that it remains only a nominal appendage to project titles, is the first item to be dropped when budgets are threatened, and the environmental consequences of that decision are ignored.

*Industry*   Industrial development and modernisation have been out of fashion in development circles for some time. On the surface this would place Allan's plea (1992; 12) for water to be transferred from uneconomic irrigation to hard currency earning / job creating industries firmly in the unsanctioned category, at least in Western development discourse. The fact that Ta'iz industry earns 275 times more capital and provides 21 times more jobs than the irrigation sector supports Allan's plea. The historical levels of industrial pollution in Ta'iz are the major drawback to this argument. However, with several of the major industries looking seriously at waste water treatment and the only treatment in the city at the moment being by industry, there is significant potential to overcome this objection. Government legislation, with monitoring and enforcement, regarding industrial pollution is still much needed, but should also be targeted at domestic waste. Urban domestic waste water accounts for around 50% more pollutants entering the surface water and groundwater than industrial waste water does. This stands in marked contrast to the development literature reporting on Ta'iz, which seems to spot the speck in the eye of industrial pollution and miss the plank of domestic pollution on the few occasions it does mention the environment. Perhaps this is because the development recommendations also want to avoid commitments to domestic waste water treatment, and they do it by pointing the finger at industry.

*Virtual water*   Closely related to the irrigation/industry debate is the issue of food self-sufficiency versus virtual water. The argument runs that widening the economic base (particularly through the development of industry) will not only provide livelihoods but will also earn the country more currency with which to buy food staples from the world market. In the case of Ta'iz, industrial development would pay for food imports even if the latter had no subsidies either from the Western exporters or from Yemen. However, the siege mentality of national food self-sufficiency remains one of the most unmoveable beliefs for Yemenis, whose hearts are in the village and the soil as much as the Israelis' were in the kibbutz forty years ago. The main source of inertial resistance to moving away from food self-sufficiency arguments is that they justify the effective subsidy of diesel prices for the land-owning ruling class, thus maintaining their income.

*Demography* Perhaps the most limiting factor for sustainable development is the population growth rate (Allan, 2000). The population problem is not simply a fixation of the environmental movement (Jewitt, 1994). Yemen is still in the steepest part of the demographic transition (Figure 3.3). The national average is 3.5% pa and the city growth rate over double that figure. At that growth rate, even if all the potable water ($<1500\mu S/cm$) were to be abstracted from all the major wadis in a 50Km radius and unaccounted for losses reduced to 20% a population of 1.1M people (by 2013) could be provided with only 30 lit/day. This abstraction would have a similar impact on irrigation to that which occurred in Wadi Al Hayma, that is, its termination. Every other factor related to water transfers mentioned up to this point in the discussion pales into insignificance if the demographic issue is not faced. However birth control, or children by choice, remains firmly in the realms of very unsanctioned discourse (Allan, 2000; chapter 4). One can only assume that before long people will seek employment elsewhere and that the pressure to migrate to other parts of the Arabian Peninsula will increase. Government could consider offering industry incentives to establish new factories on the coast and use their returns to water to pay for desalination.

*Tax, corruption, law and order* Morton (1994) notes that the failure to collect tax, deal with corruption and enforce law and order are root causes of the poverty of Eastern nations, and yet that these issues are not allowed to be challenged by the development process. These failures are characteristic of the politically powerful (Ostrom, 1999; 199) of weak states, and have a similar debilitating affect in Yemen. For instance, Saqqaf (1985) reports that only 38% of tax due is collected. Although the World Bank has raised the issues of tax collection and corruption (Hildyard, 1998; 43), since Al-Hamdi there has not been much sign of progress in Yemen.

*Measures of economic development: GDP, remittances, and national budgets* Whether GDP is an accurate measure of economic development is questioned by World Wildlife Fund (1996). Assessing Yemen's economic development is also difficult. There is a significant smuggling economy and over 60% of foreign currency transactions never enter the banking system (Yemen Times; 12/6/95). Prior to the Gulf War (1990), the scale of remittances not banked was huge. Much of these funds were invested rather in irrigation and houses in the villages and towns and in setting up small businesses. From the urban household water use survey (Handley, 1999a) it was possible to estimate the value of the land, houses, vehicles and a few major household items belonging to the Ta'iz householders. The average value of these things alone, not including any other items such as business assets, was $62,000/household if land in the village was included and $47,000/household if it was not. When it is also considered that 40-50% of the budget is spent on defence (Saqqaf; 1985, Yemen Times;17/4/95) it must be questioned whether aid to Yemen is equitable for the rest of the world, that is, whether other nations might not be more worthy recipients.

*Education* Dixon and Hamilton (1996) found the human asset capital stock of countries to directly correlate with the mean years of education per capita. Investment in education is no doubt laudable, and Milroy's hypothesis that education was a cause of terrace degradation (1994) is rejected (emigration was a much more likely cause). In the search for water for Ta'iz, the failure to conduct pumping tests correctly raises the question of the role of education in development. Although this particular example contributes directly to the on-going water problem, it is symptomatic of institutional failure resulting from indigenous educational weaknesses. It is suggested that the potential for institutional reform and improved social adaptive capacity are ultimately limited by the quality of the staff within those institutions, which in turn is largely determined by the educational system to which the staff have been exposed.

The Yemeni education system should be examined, particularly if development funding in the water sector is to be channelled through education and training. Firstly the rote memory system mentality which is developed as a first stage of education in the Qur'anic schools seems to permeate to the highest levels of education. This results in a lack of ability to apply knowledge on the basis of principles. Secondly the extent of cheating in exams raises questions regarding the value of qualifications. This leads on to the rarely perceived problem of having a higher respect for qualifications over experience (contrast Manzungu, 1999; 159 and Berkoff, 1994; 47) than occurs in the West. Fourthly, having invested in the education of tomorrow's forward thinking hydraulic scientific community, some of the most able emigrate (Morton's brain drain, 1994; 47) whilst others find there are no jobs through which to turn qualifications into experience (ibid.). Fifthly, the educational content imparted by the West to the East is becoming so high-tech that it is creating a screen-glued elite of technocrats who disdain dirty grass roots field work and indigenous knowledge and become even more detached from the problems than did the old style "technology transfer" development method.

*The development process* In terms of the layered model of water shortage causation (chapter one), the focus of development seems to have moved progressively closer to the core, through successive emphases on the physical supply of water, macro-economics and now institutional issues. However, it must be questioned whether Eastern governments will allow development agencies to encroach any further, that is, whether legal frameworks, political interests or, ultimately inherent belief systems will be allowed to be challenged.

Faced with 25-30 l/c/d of low grade water in Ta'iz and the consistent plea for spare parts for rural supply schemes, it would appear that most consumers need Serageldin's old agenda (1994) of household services provision. The existence of the World Bank's Ta'iz Water Supply and Sanitation Project also recognises this need. Although needful, Allan considers that 'comprehensive pipe water schemes can only be introduced sustainably into strong and diverse economies' where the institutions can 'deliver a flow of funds to operate and sustain them' (2000; chapter 3). Because this does not yet exist in Yemen and because urbanisation continues apace, it might be suggested that, in pursuing piped water schemes, development is only creating the need it seeks to meet. This raises perhaps the most unsanctioned topic for discourse of all; the development process itself.

*Knee-jerk emergencies and windows of opportunity for what?* Winpenny (1994; 83) commented that short term development measures introduced, for example in an emergency, might have a strong immediate, positive impact which would be likely to tail off. Such has been the case in Ta'iz, where lurching from one "emergency" to the next has been the norm. With the provision of inadequate quality water once every three weeks under "normal" circumstances, it could be contested that water provision in Ta'iz is in a continual state of emergency. However, the coincidence of a low rainfall year and a political opportunist in power tends to result in an "emergency", or "window of opportunity" (Kingdon, 1984). Thus it is difficult to discern whether the window is an opportunity for economic/scientific common sense to prevail, or is a "window of political opportunity". To date, the emergencies seem to have resulted in panic drilling, that is, jobs and money for whoever can get involved, and the irrational nature of handling the emergency ensures there will be another one in a few year's time. The short-termist attitude which seems to accompany the emergencies has resulted in a new scheme of some kind every seven or eight years on average, a point which some development agencies seem to ignore when planning the next one.

*State – development agency symbiosis*  Migdal (1988; 21) links the role and effectiveness of the state domestically to its place in the world of states. This especially proves true in the Ta'iz development arena, if development agencies align themselves with central government, which they have to work through at least to get their residence permits, giving the government legitimacy (Vincent, 1991; 210). Making field visits in one of the President's layla 'alawis both confirms the allegiance and dangles the dollars of the World Bank before the locals, winning at least superficial allegiance to those associated. All political levels from the local shayx to the President and his party will want to be seen to be a vital part of the mechanism that results in development funding ending up in the locality. How much slips through the nets of those involved and arrives at the intended destination is another matter, but as long as a large enough amount does, then both political and financial mileage can be made by those involved, and aid can remain part of the problem as well as part of the solution (Hildyard, 1998; 48).

*Jobs for the boys*  Lastly, it must be questioned who gets involved on the donor side and why. In the pursuit of more appropriate development, Western academia seem happy to keep moving the goalposts. The fact that this provides opportunity for publications and attracting research grants or "academic consumables" (Yoshida, 1999) is not coincidental (Bryant, 1998; 13). It is curious that on the basis of lucrative data intensive studies (Chambers and Carruthers, 1986; 2), consultants propose schemes that can fail (e.g. Al Hayma) without being sued for getting their numbers wrong. Perhaps this is because the water runs out five years later instead of two weeks later as in the dewatering sector. Even in the name of institutional capacity building there has still not been a well-conducted pumping test in the Ta'iz area. It is not surprising that development workers give up, when they know their advice will be ignored (Morton, 1994, Allan, 2000; chapter 4), the infrastructure will not be maintained whatever is done with the institutions, and funds and equipment will continue to find their way into the wrong hands.

*Summary*   This section has demonstrated that deeply held beliefs, conscious and sub-conscious, contribute to determining what can be discussed and what cannot in the field of water allocation. Many facets of Yemeni life impinge on the water allocation nexus and the degree to which discourse on them is sanctioned or unsanctioned determines what contribution they can make, if any, to keeping the population fed and watered.

The inability to sanction discourse on existing/potential water shortage causes/solutions is not only a Yemeni problem. Western development discourse can also exclude key issues. The next section examines some of the limitations of Western sustainable development models in the Ta'iz context.

## Equitably, Environmentally and Economically Sustainable Development

The history of water allocation and use in Ta'iz provides a useful context against which to test the relevance of the different schools of thought regarding the theory and practice of sustainable development.

*Introduction: Getting Things in Perspective: Environment and Equity*

In chapter two, the extent of degradation of the water environment was investigated and it was demonstrated that water levels had declined between Habir and Al Burayhi, in Hawban, Wadi Hidran and beneath the city of Ta'iz. The result has been declining wadi flows, declining well yields and in many instances the drying up of wells. The area affected accounts for around one third of the total stream / groundwater irrigated land of the Upper Wadi Rasyan catchment that was irrigated at the height of groundwater development in 1985. A further quarter of this total has been polluted directly by urban domestic and industrial waste water. The two portions do not coincide, resulting in over half the stream / groundwater irrigated land becoming environmentally degraded.

Chambers (1983) identifies a deprivation trap into which the powerless, vulnerable, physically weak, poor and isolated downwardly spiral. He notes that the rural and urban poor are particularly susceptible to this process and Berkoff (1994) also suggests that underprovision of water and sanitation is typically skewed towards the urban poor. The following summary demonstrates that both these observations are applicable to Ta'iz. In chapter three, the provision of water and sanitation services, and also the contribution water makes towards providing livelihoods, were examined.

In these contexts, it is worth raising the equity related questions of whose livelihoods and whose quality of living are affected by either water allocation practices or degradation of the water environment.

*Water and sanitation services* In over 80% of rural areas, women continue to carry water distances averaging 1.7 Km and spend two hours per day doing it. In the city, the poorer 60% of the population spend 50% longer fetching water (around 10 hours per week) and twice as many poorer children are involved compared with the wealthier 40% of the population. The poorer 60% spend an average of 6% of their income on water compared with 2% for the wealthier 40%. Two-thirds of the urban community cannot afford tankers and one-quarter have to drink water of lower quality than recommended by WHO because they cannot afford treated drinking water. Water tankers are more expensive for poorer families, and those recently moved into the city, who have to build their houses at the city edges face higher tanker charges, have less access to public water supply and sanitation services. The wealthy also benefit more from the subsidised public supply and sanitation because they have far more water consuming facilities and gardens. Poor sanitation facilities are also beginning to affect the growing rural population and those downstream of the city and the factories are forced to use polluted water, and know it.

*Livelihoods* The agricultural sector has been especially affected by water level declines. The wealthier, more powerful landowners can deepen wells to chase the retreating resource, hence stealing it from under the feet of the poorer ones. Principal-agent problems prevail in negotiations for compensation for water reallocation and 3$^{rd}$ party interests tend to be underprotected. The pollution of groundwater and soil in once fertile wadis has affected all downstream of the city, and poorer farmers adjacent to industrial plants have had to abandon land or sell cheaply. Although returns to water in terms of livelihood provision are far better in industry than in agriculture, pay differentials are rather inequitable in the sector.

*Choosing a model: the sustainable development triangle* Current discourse on development appears to revolve around three basic aims; economic progress (or growth measured in some manner such as per capita GDP), equity provision (a more even distribution of goods, or, less often mentioned, "bads") and environmental protection (from resource depletion or, taking it further, even resource reconstruction). These aims can be envisaged as separate axes or poles (Figure 4.5).

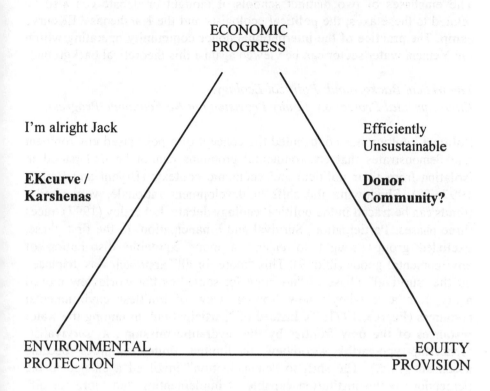

**Figure 4.5 The unequilateral triangle of sustainable development**

The emphases of two distinct schools of thought or debate can also be related to these axes; the political ecologists and the Karshenas / EKcurve camp. The practice of the international donor community operating within the Yemeni water sector can be viewed against this theoretical background.

### Theoretical Background: Political Ecology:
### Environmental Protection, Equity Provision but No Economic Progress

Political Ecology has contributed the concept of a politicised environment and demonstrates that environmental problems cannot be understood in isolation from their political and economic contexts (Bryant and Bailey, 1997; 28). Shadowing the shift in development rationale, very similar trends can be traced in the political ecology debate. Eckersley (1997) traces three phases; Participation, Survival and Emancipation. In the first phase, excluded groups sought to ensure a more equitable distribution of environmental goods (ibid; 9). This "more for all" approach was displaced by the "survival" phase of "no more for some" as the worldview moved away from a 'cowboy's new frontier' view of limitless environmental resources (Pearce, 1993; 2). Instead of "participation" in taming the water resources of the new frontier by the 'hydraulic mission', a 'survivalist' 'spaceship-type-earth' worldview of limited finite resources became ascendant (ibid.). The shift to "emancipation" involved a change in the perception of the institution capable of implementing "no more for all" from Leviathan top-down enforcement (Lam, 1994; 15) to bottom-up self-applied restriction on consumption by an conserver society aware of its responsibilities to future generations. The emancipation stage has also questioned the very notion of material progress (Eckersley, 1997; 17) and because Western rates of consumption are so high, it is implicitly suggesting "less for some". Although the participation stage embraced economic progress, the survivalist stage viewed economic progress and environmental protection as opposites and the emancipation stage completed their antithesis.

### The Karshenas / EKcurve Debate:
### Economic Progress and Perhaps Environmental Protection but No Equity Provision

The current stance of the political ecologists, or at least of the deep ecologists, assumes that economic progress is simply incompatible with protecting the environment. This assumption is directly questioned by Karshenas (1992; iii and 22). In essence following the Environmental

Kuznets Curve (EKC) model, Karshenas observes 'an unmistakable complementary relation between employment generation and environmental generation' so that 'the more advanced a country and the higher its technological level, the cleaner its environment becomes' (ibid.).

On the basis of Western post-industrial nations Karshenas' hypothesis states that it is possible to reconstruct a cleaner environment, but that it is first necessary to undergo some environmental degradation (theoretical curve, Figure 4.6) as the economy is strengthened. He contrasts the political ecology view that economic progress and environmental protection are mutually exclusive and "traded-off" against each other versus his own view that the two can be "complementary". Others strongly contest the theory that the history of development in Western nations supports the validity of this view (World Wildlife Fund, 1996). Many countries do not seem to have "turned the curve", that is, their environments are still deteriorating. This forms one major objection to the Karshenas hypothesis. However, the fact that he places environment on one axis and economic growth on an axis perpendicular to it, rather than the political ecologists' polarisation of these variables, at least gives some hope, if not scope, for the development that some economies (such as Yemen's) need. However constructing axes without data indicating that there is a turn on the "Ekcurve" does not prove anything.

Eastern countries in particular show no signs of turning the curve, however, Karshenas specifically addresses this issue by differentiating between Western and Eastern development. The latter is hampered by what he terms 'forced environmental degradation' observed as an 'unambiguous complementarity between economic underdevelopment and environmental degradation' (ibid.; 14). The cause is identified as a 'low rate of increase of man-made capital stock, technological backwardness and stagnation, combined with a population growth which eats into the natural capital stock' (ibid.). In this situation, he prescribes a rate of economic growth which should be 'sufficient to cater for the basic needs of the population' (ibid.; 10), and on this basis defines sustainable development as a 'feasible'... 'minimum socially desired rate of long-term growth'. "Long-term" means tolerating the shorter-term environmental deterioration now in anticipation of turning the curve in the future. The curve may be the donkey's carrot always lying tantalisingly ahead and maintained there by population growth.

Catering for the basic needs of the population is a rather ambiguous term. Where do basic water needs stop? Drink, food, domestic water, livelihoods? Each of these stages needs more water.

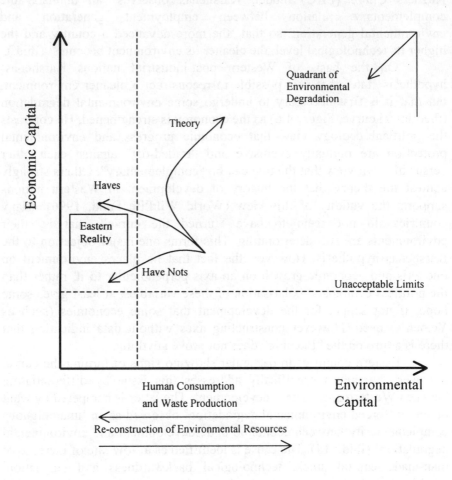

**Figure 4.6   EKcurve / Karshenas views of sustainable development vs. eastern reality**

At this point it is worth evaluating Ta'iz in terms of Karshenas and vice versa. Firstly, the water situation in Ta'iz is unsustainable. The 1995 water crisis proved that the system could not maintain its productivity when subject to stress or shock (Conway's definition of unsustainability, 1986).

One would expect Ta'iz to fit Karshenas' definition of forced environmental degradation (above). However, two different situations are apparent: the rural areas where environmental degradation is measured primarily in terms of declining water levels, and the city together with its downstream impacted areas which are characterised by polluted surface and groundwaters. In the rural areas, rather than environmental degradation being a result of the pursuit of maximising agricultural production to feed an increasing population (Karshenas; 1992; 23), more and more irrigation has been used for growing cash crops such as qat, whilst virtual water in imported wheat has made up the deficit and fed the growing population. This is not forced environmental degradation but chosen environmental degradation by Hardin's selfish hedonists (1968). In the city, urbanisation has bred urbanisation and concentrated population, resulting in a deteriorating environment of increased pollution. The fact that increasing pollution remains unchecked is considered to be primarily due to the absence of institutional and legal frameworks to make the polluter pay. This is a reflection of Yemen's limited (social) institutional adaptive capacity (Ohlsson, 1999 and Turton, 1999c).

Karshenas asserts that dealing with the root cause of population increase is less amenable to policy intervention than are technological change and employment provision (1992; 24). Technological change has certainly contributed to the cause of environmental degradation in Ta'iz with the development of tubewells and pumps. In contrast, on the basis of the huge differences between agriculture and industry in returns to water, it can be emphatically stated that technological change in the form of a growing industrial base, is a more viable solution to the problems of "forced environmental degradation", since:

1.  Industry addresses the problems of a stagnant economy, underemployment, technological backwardness and low levels of man-made capital (ibid.) by providing employment which has higher returns to water and technological levels than other existing forms of employment. It also employs more people and creates more wealth in absolute terms than the alternative livelihoods and increases the level of infrastructure (man-made capital).

2.      Industry feeds the population in an indirect way by underpinning a stronger and more diverse economy (Allan, 2000; chapter 4) which could facilitate the purchase of grains from the world market.

If handled badly, however, industrial development causes, and in Ta'iz has caused, further environmental degradation, just as the political ecologists suggest. If handled well, with enforced regulation, industry can be made to return used water to the environment at acceptable quality, and hence follow the Karshenas model. Indeed there are many signs that at least the major industries of Ta'iz are willing to adequately treat waste water without being forced to. The political ecologists in rejecting industry (the very soul of capitalism, Atkinson, 1991; 5) may in fact have thrown out the baby with the dirty bath water.

Economic signals and preferences emanating from central government do not seem to encourage the Ta'iz domestic sector to treat its waste water. There are therefore two problems. One problem is the handling by government (with or without legislation) of both industrial water users and especially domestic water suppliers. It is precisely this point that Ohlsson's (1999) and Turton's (1999c) comments regarding "social adaptive capacity" of politicians and governments addresses. The observations regarding the failure to pass a water law and the failure of government to enforce law begs the question whether low "social adaptive capacity" is synonymous with a weak state/strong society such as Yemen. The second problem is that the donors have also largely failed to acknowledge the contribution industry makes to intersectoral allocatively efficient water use and to the solution of environmental degradation. Relative to support for irrigation, the donors have provided little if any encouragement towards industrial development, and government and donor incentives to industry to treat waste are lacking. Together with their emphasis on irrigation, this suggests they too may be lacking the "social adaptive capacity" to embrace environmental issues. The two problems of inadequate legislation / enforcement and ignoring the contribution industry could make to cleaning up its act together with the role the private sector plays in providing water suggests the authoritarian state and free-market allies of the emancipatory theorists should not have been totally rejected (Eckersley, 1992; 28).

Finally, there are two facets of the Karshenas model which remain unclear. Regarding the reconstruction of environmental capital there is the problem of "historical datum".

What abstraction rates should we be aiming at? Allan (1994a; 3) asks whether past patterns of use should be taken into account. In Al

Hayma, donors, government and farmers alike talk about consumption levels of 15 years ago, when groundwater abstraction for irrigation was at its height. Surely, for environmental recovery in Ta'iz, levels of 30 years ago would be more appropriate? If the Hittite race still existed, the Palestinian – Israeli argument over the West Bank might be solved – or at least have another contender from a few thousand years earlier. The historical datum issue also isolates the fundamental problem for Ta'iz. Water scarcity is such that it is impossible to return to any previous levels of agricultural water use because the population has increased so much in the mean time. We cannot reproduce the past because of today's population level. Virtual water and modern medicine to a large extent mitigate against the ancient means of population control of famine and plague, leaving war as the alternative. If we accept the level of population growth and seek to feed and water it, then we must look to technological change for a solution (Karshenas, 1992; 24). Indeed, in Ta'iz to date, technological development in the form of industry has done most to meet the population's livelihood needs.

The second, and more serious flaw in the Karshenas model, is that although the model finds a harmony for environmental protection and economic development, it ignores equity. Falkenmark and Lundqvist (1995) and Laird (1991; 17) also raise the issue of economic progress and equity provision. In particular the population of Ta'iz demonstrates that the haves can invest the fruit of their economic progress in water transfers (by tanker or whatever) and create their mini-paradise, whilst the have-nots downstream experience declining water availability and quality and rising water prices. This phenomenon reflects the power asymmetries of actors controlling the environments of others which Redclift points out (1987; 49) to be a particularly Eastern feature. The Karshenas/EKcurves may hold true for the haves, but for the have-nots it moves towards the minimum environmental capital and minimum economic well-being quadrant of the graph (Figure 4.6). World Wildlife Fund (1996; 25) says 'if the benefits from economic growth accrue to a small minority of the population, only this small minority will be in the position to demand a cleaner environment'. Such is the case in Ta'iz.

There are also indications in Ta'iz that the EKcurve haves/have-nots split is not just local. Whether Proctor and Gamble, Shell, Rothmans, Crown Paints etc will be as rigorous upon themselves regarding their impact on the environment of the Ta'iz area as they have to be in their home countries is not yet apparent. Wealthier countries may prove to be exporting the "unsustainability" effects to Ta'iz when considered in terms of the negative environmental impact they are making (World Wildlife Fund; 1996; 10).

*Can You Have Your Water and Drink It? Donor Agency Practice: Economic Progress and Equity Provision, but Only Lip-Service to Environmental Protection*

Since initial development drives for modernism and industrial and agricultural development, environmental issues have come to the fore, contributing to adding the word "sustainable" to the development debate. Development then shifted to a bottom-up participatory / stakeholder institutional emphasis. However the development community does not seem to be as radical as the deep ecologists, and still assumes material progress where there is "all for some and some for all", at least in the water sector (Delft, 1991). Donor activity in Ta'iz, even today, comprises dams and wells for irrigation in the compensation packages for rural-urban water transfers. Irrigation productive efficiency measures are proposed, despite the fact that these will only produce "efficient unsustainability", and treatment of sewage continues to be dropped from the agenda. The reality of donor activity causes one to question whether the "environmental bit" is really just an 'add on' of secondary importance (Redclift, 1987; 14).

Within the sustainable development debate three camps, and their biases, have been identified (Figure 4.5) and can be related to the Hippocritical-style cultural theorist approach (Thompson, 1995; 32). Political (deep) ecology promotes environmental protection at the expense of economic progress. For a region with the economic limitations of Ta'iz, that expense may be excessive. Karshenas (1992) tries to reconcile economic progress and environmental protection, but at the same time ignores the inequitable distribution of both the environmental and economic goods and bads. Whether there will be a turn in the curve for the majority of the population remains to be seen. To date, industry has been the technological innovation to address the problems of a stagnant economy, underemployment, technological backwardness and low levels of man-made capital as well as provide food by facilitating an economy that can pay for virtual water. Perhaps economic progress and environmental enhancement can go together (Serageldin; 1994; 94) but we still need to clean up in order to survive rather than postponing the clean up until we are surviving more comfortably. To initiate the clean up some economic signals from government and donors would be welcome. Donor agencies appear to shy away from environmental protection, despite paying it lip-service.

## Conclusion: Individual, Economic, Institutional and Political Adaptive Capacity to Water Shortage and The Integrated Holistic Model

The ability to innovate in the face of complex challenges such as the water shortage of Ta'iz is a reflection of social adaptive capacity (Turton, 1999b; 10, from Ohlsson, 1998 and 1999). Allan (1999a; 3) finds the required innovation in water policy reform, needed to improve water use efficiency via institutional reform lacking in "weak states". The failure of Yemeni government institutions to cope with water shortage defines its "weak state" status. However, this chapter has demonstrated that political self-interest at all scales, from shayx to central government ministries, and above, have also been a fundamental obstacle to water use efficiency. In contrast, there are also several indications of significant social adaptive capacity to water shortage in Yemen.

Firstly, there is the enormous adaptive capacity demonstrated by the change in behavioural patterns of individuals in response to water shortage in the city. In this instance the extent of adaptation is the greatest, the scale is the smallest (individual households) and the incentives to adapt are the strongest (survivalist, need-driven adaptation). Secondly, the preparedness of individuals to sell water to tankers and industry with the formation of markets and the development of private rural water supply schemes are forms of economic response to water shortage and reflect economic adaptive capacity. Thirdly, the establishment of water user associations and water related initiatives (such as irrigation co-operatives and community supply schemes) in some rural areas represent a more communal response to water shortage or institutional adaptive capacity. Fourthly, the negotiations of community representatives with government over water transfers such as has occurred in Al Hayma and Habir, is a form of political adaptive capacity to an external shortage. The degree to which the community or just the representatives benefit from the outcome may reflect the democracy of the process, but in either case there was capacity to adapt. These four examples form a sequence of increasing institutional scale and, necessary to that increase, an increase in political profile and, not incidentally, volume of water involved.

The term "social" adaptive capacity is intended to cover a range of contributing factors including economic (Turton, 1999c; 13), socio-political (Allan, 1999b) and institutional (Allan, 1999a; 3). Discussion is typically in the context of central government initiatives whereas the Ta'iz examples of adaptation to water shortage mentioned above are all facilitated by individuals apart from central government. (Even in the fourth example water transfer is facilitated by local shayxs not central government.) The solution to this contradiction lies in Turton's assertion that social adaptive capacity is a function of government legitimacy (1999c; 13). In Ta'iz, and probably many other strong society – weak states, the local indigenous, informal mode of government is simply more legitimate than more centralised forms. In conclusion, there can be some measure of social adaptive capacity in a weak state providing the process of allocative reform starts as a bottom-up process at the local "informal" level of real politik and uses those equally indigenous institutional and legal frameworks in which it has traditionally operated. Central government will have to be politically more transparent if it is to earn the trust to enable it to play a more significant role in water transfers as negotiator or even buyer, in the form of NWSA, in the future. In order to facilitate institutional change it has been suggested that central government should not become a bargainer at all (Steenbergen, 1996; 204).

In the context of Yemen, the holistic model of layered causation in the allocation of water proposed in the introduction:

Political => Legal => Institutional => Economic =><= Social => <= Physical

fails at a number of points. In the strong society – weak state of Yemen, it is discovered that state-originated legal frameworks and institutions become irrelevant at the local level, the very level at which water reallocation is most needed. Instead, local and/or traditional legal and institutional arrangements are the main vehicles of allocation. Indeed the political interests which underlie or work through those arrangements are themselves a reflection of specific local individuals rather than any Western party-political model. Perhaps so pervasive and obvious they can be overlooked are the belief systems which drive the political interests and determine the political possibilities. The local political reality is embodied in the shayx, who is also the dominant institutional reality. This duality of function can be traced to an extent up to national level ('the state is part of the tribes', The President, in Dresch, 1989; 7). Alongside this traditional structure is a fledgling bureaucracy under a civil code created by the parliamentary process. This is the more typical Western model described in the introduction and shown above. It is suggested that in Yemen there lies a spectrum between these two models (Figure 4.7). Opportunity for principal- agent abuses are greatest in the "law-less" middle. Yemen may also be considered as in a transition from the Eastern to the Western model. It must be hoped that in attempting to transfer, Yemen does not get caught in the middle. Finally, the lack of action on protecting the water environment suggests that this aspect only exists in rhetoric.

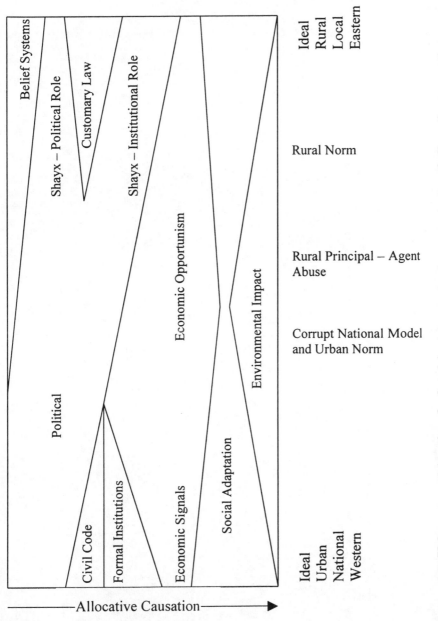

**Figure 4.7 Urban – rural / national – local / western – eastern spectrum of allocative causation**

# 5  Conclusions

A summary of the problems causing the water shortage in Ta'iz points clearly to the need for an integrated holistic approach to water resources management. By the end of the 1990's, it had become clear that "integration" was not an easy process; contention is the norm in user relations over water (Allan et al, 1999). Integrated water resource management has proved to be an extremely alien concept, though vital to the understanding of the Ta'iz story of combined resource and management failure. The history of the shortage so far necessarily ends with the completion of fieldwork in the Upper Rasyan Catchment in 1998. In concluding, some consideration is given to the significance of the shortage and the lessons derived from it beyond these temporal and spatial points.

**Key Issues**

*Water Reallocation: The Need for Economic Diversity*

Perhaps the most important point arising from this study is the small amount of water that could be made available for reallocation. A mere 30Mm³/yr is the best estimate of renewable groundwater from the Upper Rasyan catchment to meet the drinking, domestic and livelihood needs of around ¾ million people. The food needs are mainly being met by a reallocation of water from overseas in the form of grain imports. Their virtual water content accounts for over three times the amount of water found in Ta'iz's renewable groundwater. Economic diversity towards allocatively more efficient uses of water provides the capacity to afford grain imports. In the debate over whether to produce cash crops or industrial output with the little water available (Otchet, 1999) the deciding factor in an economist inspired world is the relative returns to water. In Ta'iz, industrial use of water brings 2300 times more income and over 300 times more jobs per m³ of water than does irrigation. The cessation of irrigation in the area is not advocated. It would not solve the problem. A reallocation of a small part of irrigation's 75% share of the renewable resource to industry would help, however. A 10% reduction in irrigation water use could produce a 250% increase in industrial use if reallocation were feasible.

## *Strong Society – Weak State Implications: The Need for Political Credibility and Institutional Appropriateness*

The reallocation of water has been feasible on a small scale through water markets but transfers from rural areas to meet the growing urban demand by government has met with armed conflict and produced significant environmental impact. The armed conflict was fostered by mistrust of government on the part of the source area communities, which in turn was brought about by abuses of their agent-principal relationship. Although such abuses also sometimes characterise local traditional indigenous power structures, these structures operate on the basis of customary law via traditional non-formal institutions which are respectively more enforceable and considered more appropriate and efficient in the local context. Political leadership at the local level is individualistic and family-based rather than tribal, and the adoption of national party-politics by those individuals is on a purely pragmatic basis. Local political leadership wins greater credibility than government because of the latter's poor track record and "distance" from the issues. The Yemeni North - South relations also play a part in this distancing. The government also fails in the role it could fulfil – in providing the right macro-economic signals to promote reallocation to higher value uses of water. Traditions of community co-operation in rural areas have, in a few instances, paved the way for the establishment of co-operatives involved in water related activities and, with some donor input, water user associations. The tendency towards local rather than government based power structures reflects and demonstrates the region is characterised by a strong society and a weak state. Brumfiel's definition of a state 'in which governmental institutions monopolise the use of legal force' (1980) suggests Yemen is not a state. To be effective, water reallocation initiatives have to operate within this non-state context.

## *Population Growth: Still on the Agenda*

If unaccounted-for-water was reduced to 20% and all the renewable groundwater in a 30km radius were brought to Ta'iz, only 30 l/c/d water could be provided in 2013 if all irrigated agriculture were to cease. The impossibility of such a plan both physically and socio-politically underlines the fact that the most onerous factor affecting the area's water needs is the city's 8% p.a. population growth rate. Any supply side development of those meagre renewable water resources is only buying time (Turton, 1999d; 1). The only other supply alternatives are desalination at over $2.5/m$^3$ and/or domestic water recycling which is unacceptable on religious grounds.

## The Sustainable Development Balance:
## Economic Progress, Environmental Protection and Equity Provision

The people of Ta'iz receive less water from the public utility than the WHO guideline and that water is of poorer quality than the WHO recommended minimum. Such circumstances demand development initiatives. Economic progress in the form of infrastructure renewal and utility reform is a necessary ingredient. The pollution of over half the catchment's groundwater and surface water and declining water levels in key aquifers demand environmental protection initiatives. The poorer 60% of the population spend more on water than the World Bank recommended maximum. The result is an average of half an hour per day queuing for free water in the city and carrying it 1.7 km in rural areas. This price is mostly paid by children and women respectively. Equity provision is therefore on the agenda. The need for a balance of all three; equity provision, economic progress and environmental protection is apparent.

## Summary

The reality of water allocation between rural source areas and urban consumers in the Ta'iz area reflects political rather than economic factors, and economic rather than water resource criteria. This hierarchy of causation results in pollution of the rural area downstream of the city. Although keen to increase supply to the city, donors seem unwilling to address the waste it creates.

Drinking and domestic water needs have been met increasingly by a combination of highly adaptive private sector supply initiatives and equally adaptive consumers forced by the shortage to reduce demand. Water shortage in Ta'iz has forced many individuals to change their water use practices. However, the modifications of demographic and religious practices demanded by the shortage are a severe challenge to an individual's belief systems, and hence, social adaptive capacity. On a larger scale, the underlying belief system of political self-interest that constrains reallocative initiatives and silences the role of virtual water is also unlikely to adapt to the shortage.

## Beyond Ta'iz

### Ta'iz Tomorrow

The observations of the Ta'iz water crisis were snapshots in an evolving situation. Whilst belief systems may change very slowly, population levels and the amount of imported grain to feed it and technology and the extent of environmental degradation it causes have changed rather quickly. Optimists hope there is a demographic transition. There is no sign of a drop in the birth rate yet. The growth in tubewell/diesel driven pump technology was fuelled financially by remittances from Saudi and the Gulf and literally by the discovery of some local oil. The environmental consequences suggest that black gold brings a mixed blessing. The problem of trying to return to the environmental condition before significant groundwater development occurred is that the damage has already been done. At current trends, population growth is likely to prevent the EKcurve being turned. Beyond the ten-year horizon, the sectoral reallocation issue becomes irrelevant because it cannot find enough water. The only solutions then lie in virtual water for staples provision, industrial development (with regulated waste disposal) for livelihood provision and, in the absence of waste water treatment, desalination for industrial and domestic needs. The latter measure would require a regional economic strength that can give one million people in the city the capacity to afford water at over $2.5/m^3$. That cost could be halved if industry moved to the coast since this would avoid the 1400m water lift and 100km pipeline needed. However, it is doubtful that the water would be affordable to most Yemenis.

Ta'iz's struggle to balance in the water allocation nexus of economic growth, environmental protection and equity provision can also be positioned in time from the vantage point of Western development models of society and the state.

In the West failure to protect the environment began to be noticed after a century of industrial modernity had damaged it (Giddens, 1990, Allan, 1999b; 2). Since then the environment issue has ascended to the point where today Western industry, in the light of perceived "risks" to the environment, has started to take a more precautionary approach to the use of natural resources (ibid.). This shift in perception and policy has been termed "reflexive modernity" (Beck, 1992). The social theory of Beck and

Giddens is relevant to the West, where Western awareness of "risks" has resulted from assimilation from the media by communities and their politicians of the "new knowledge".

Where does Ta'iz lie within this scheme? Allan (1999b; 3) gives four reasons why the East will find it difficult adapting to the message of the "new water knowledge" inspired from principles of economic efficiency and environmental sustainability of which water reallocation and birth control are a part:

1.  Deeply held beliefs and expectations about water that contradict the message.
2.  Lack of susceptibility to the notions of environmental risk.
3.  Lack of political, social, economic and technical capacities to respond to the message.
4.  Awareness by the government policy makers of the economic and technical impediments and, more importantly, political prices to be paid in applying the message.

Although all four points are true of Yemen, they assume central government initiatives in water policy reform will have an impact. The limited institutional capabilities of central government and the limit of its effective rule at, or, if kidnappings are an indicator, even before, the San'a ringroad, indicate that the processes of state building involving the development of a bureaucracy and the disarmament of the people (Weber, 1978) have not yet taken place. Yemen lies at least as far from applying the "new knowledge" than any other polity in the East. Besides the incongruity of encompassing Eastern state development within Western models, is the fact that those models are themselves part of an evolving literature forever moving the goalposts where today's solution for yesterday's problem is the cause of tomorrow's problem.

*Ta'iz Elsewhere*

How relevant is the example of Ta'iz to other situations, or, put another way, how typical is Ta'iz? If only six main features could be chosen to describe Ta'iz they would be:

1.  The Problem - The non-availability of water is extreme (around 25 l/c/d).
2.  The Main Cause - The population growth rate is particularly high (around 8% p.a.).

3.　　　An Evaporating Hope - The annual rainfall though slightly greater than London's (around 600mm/yr) is still inadequate.

4.　　　The Current Food Solution - The population is highly dependent on virtual water (equivalent to around 75% of all the water used in all sectors) and unaware of its dependence.

5.　　　A Short-term Solution - reallocation - The proportions of industrial and domestic sector water use are small at only 2% and 4% of total water use, respectively.

6.　　　A Potential Long-term Solution - desalination – is likely to prove too costly.

It is hoped that the extremity of the Ta'iz predicament makes Ta'iz sufficiently different from other cities that its experience remains only a slim risk to them rather than a real threat. The lessons are still worth learning.

Water stress in Ta'iz has suppressed the demand for water, nullifying the potential contribution of demand management methods. The population distribution within Yemen (or for that matter, the Arabian peninsula) approximately matches the rainfall distribution. Yemen, and particularly the regions of Ibb and Ta'iz receive the most rain and are the most populous. Theoretically, there is no major maldistribution of people relative to water, and although there is plenty of rainfall, it occurs over too small an area and evaporates too fast, resulting in little runoff. Within the context of such natural shortage relative to the population, human activity has greatly aggravated the problem. For the time being, virtual water provides the silent solution to the food requirement whilst Yemen and cheap grain still have access to the world market. Although this could free up water for reallocation to higher value uses, currently only intrasectoral transfer to qat is evident in any significant quantity. Without a demographic transition, even the reallocation issue is only a short-term solution exacerbated the problem of unchecked pollution. If pursued, the push-pull of livelihood provision by developing industry at the coast could be a good, though costly, initiative for meeting demand (and a bad one for the voiceless corals).

## Holism: Our Man in Habir

This study sought to demonstrate that water allocation 'has too many component parts to be understandable if one limits oneself to one of the many established academic domains' (Hajer, 1995; 2). In the light of the lessons from the water situation in Ta'iz, it is suggested that a 'poverty of disciplinary narrowness' (Atkinson, 1991; 21) would prevent an analysis adequate to address the problems and solutions of water allocation. Although successive physical, economic, institutional, legal, political and belief system layers contribute increasing complexity to an understanding of the water problems, they do provide a fuller one. A succession of experts would provide an academically sounder set of reports, however, the gaps between the disciplines might bring incoherence to the message.

*       *       *

So it seemed as we returned from our field visit to the farmer in Habir on World Water Day. The encounter had a specific spatial and temporal context. The farmer irrigated his crop within a few hundred yards of the most productive exploration borehole in Habir. Permission to drill had taken almost a decade of negotiations, compensation, and bloodshed because the government had previously "stolen" the water from the aquifer and farmers immediately downstream. Government misuse of information and corruption, and opportunism by the locally powerful were all involved. The borehole was still not connected to the Ta'iz pipeline. The farmer, unlike any of the visiting experts, was well aware of all of this. Although his water came from an entirely different source than that of the drilled well, the arrival of "donor agents" and "sustainable developers", in an expensive vehicle largely determined the farmer's tactics.

On our return from the field we sat down to eat (some virtual water) and drink (some of the locally privately purified water) with the Governor. Before the meal we washed our hands in water brought by private tanker from a (polluted) wadi and stored in locally made tanks. The complete absence of a single drop of "government" water on the premises reflected the water shortage described in this study. The fact that neither we, nor the rest of the 400,000 inhabitants either died of starvation or thirst that day reflects the adaptive capacity of a society left to its own devices. (Some might have had water borne tummy bugs or malaria and some babies may well have died from dehydration from diarrohea.) It is difficult to understand these watery realities without an $H_2$Olistic approach.

# Photographs

**Photo 1  Domestic use of untreated sewage**

**Photo 2  Inequitable Water Allocation**
Foreground: irrigated qat belonging to the shayx of Lower Al Hayma
Background: rainfed land of surrounding farmers

**Photo 3  Artesian flow from fault-bounded Tawilah Sandstone**
Right foreground, dry field
Left foreground, field under 15cm water from natural artesian flow
Right background, volcanics;    Left background, Tawilah Sandstone
Lower right, pipeline pumping from Tawilah Sandstone outcrop to the
"Western scheme"

**Photo 4 Environmental degradation**
Foreground, factory waste water lagoon; middle distance, Ta'iz dump

# Bibliography

Abbott, P.F. and Tabony, R.C. (1985), 'The estimation of humidity parameters', *Meteorological Magazine*, 114, pp.49-55.

Adel, K.S. (1986), Hydrogeological report of Al-Haima wellfield, Taiz. Field visit, 1/10/85-30/11/85, NWSA, San'a.

Agnew, C.T. (1982), 'Water availability and the development of rainfed agriculture in south-west Africa', *Trans. Inst. Br. Geogr.* N.S. 7, pp.419-457.

Alderwish, A.M. (1995), 'Estimation of groundwater recharge to aquifers of Sana'a Basin, Yemen', Ph.D. Dissertation, University College, London, 312 pp.

Allan, J.A. (1992), 'Striking the right 'price' for water: achieving harmony between basic human need, available resources and commercial viability', *Water in the Middle East conference*, London, 19-20 January 1992.

Allan, J.A. (1994a), 'A transition in the political economy of water and the environment in Israel-Palestine', *Israeli-Palestinian workshop on joint management of shared aquifers*, Jerusalem, 27 Nov-2 December 1994.

Allan, J.A. (1994b), 'Overall perspectives on countries and regions', in P. Rogers, P. Lydon (eds), *Water in the Arab World: perspectives and progress*, Harvard University Press, Cambridge, pp.65-100.

Allan, J.A. (1996a), 'The political economy of water: reasons for optimism but long term caution', in J.A. Allan (ed), *Water, Peace and the Middle East. Negotiating Resources in the Jordan Basin,* Tauris Academic Studies, London, pp.75-119.

Allan, J.A. (1996b), 'Water security policies and global systems for water scarce regions', *International Commission on Irrigation and Drainage*, Cairo, Special Session R7.

Allan, J.A. (1996c), 'Water use and development in arid regions: environment, economic development and water resource politics and policy', in *Rivers and international water courses, Review of European Community and International Environment Law*, London, pp.107-115.

Allan, J.A. (1997), "Virtual water': a long term solution for water short Middle Eastern economies', 1997 *British Association Festival of Science Lecture*.

Allan, J.A. (1998), 'Moving water to satisfy uneven global needs: 'trading' water as an alternative to engineering it', *ICID Journal*, 1998, Vol. 47, No 2, pp.1-8.

Allan, J.A. (1999a), 'Water in international systems: a risk society analysis of regional problemsheds and global hydrologies', *SOAS Water Issues*, Occasional Paper 22.

Allan, J.A. (1999b), 'Facilitating remedies to water scarcity: beliefs, knowledge and action', *Global Water Partnership*, Paper 1.

Allan, J.A. (2000), 'The water question in the Middle East; the second best works', London: Tauris Academic Publications.

Allan, J.A., Karshenas, M. (1996), 'Managing environmental capital: the case of water in Israel, Jordan, the West Bank and Gaza, 1947 to 1995', in J.A. Allan (ed), *Water, Peace and the Middle East. Negotiating Resources in the Jordan Basin*, Tauris Academic Studies, London, pp.121-133.

Allan, J.A., Mohtadullah, K., Hall, A. (1999), 'The role of river basin management in the Vision process and Framework for Action up to now', *Keynote paper The Hague workshop on River Basin Management*.

Allen, R.G., Pereira, L.S., Raes, D., Smith, M. (1998), 'Crop evapotranspiration. Guidelines for computing crop water requirements', FAO Irrigation and Drainage Paper 56, Rome, 298 pp.

Ansell, C. (1980), 'Domestic water use in a subdistrict of Mahweit province. (Applied Research Report No.2)', *American Save the Children*, Yemen, 16 pp.

Atkinson, A. (1991), *Principles of Political Ecology*, Bellhaven, London.

Bamatraf, A.R.M. (1987), 'Supplemental irrigation in Yemen Arab Republic (YAR)', Chapter 29, in Perrier, E.R. and Salkini, A.B. (eds), *Supplemental Irrigation in the Near East and North Africa*, ICARDA and FAO conference, 7-9 December 1987, Rabat, Morocco, pp.561-598.

Beck, U. (1992), 'From industrial to risk society', *Theory, Culture and Society*, Vol. 9, pp. 97-123.

Beck, U. (1999), 'What is a 'risk (society)'?', *Prometheus*, Vol. 1.1, Winter 1999, pp. 75-79.

Becker, N., Zeitouni, N., Schechter, M. (1996), 'Reallocating water resources in the Middle East through market mechanisms', *Water Resources Development* 12, pp.17-32.

Berhardt, C., Griffin, R., Hawkins, R., Hendricks, D., Norvelle, M. (1980), 'Water policy initiatives for Yemen', Recommendation by CID Water Team, Executive Summary, (NWSA, San'a).

Berkoff, J. (1994), *A strategy for managing water in the Middle East and North Africa* ed. World Bank, Washington, 72 pp.

Blaikie, P.M., Brookfield, H. (1987), *Land degradation and society*, London, Methuen, 296pp.

du Bois, F. (1992), 'Regulating the competitive use of fresh water resources: problems of allocation and exploitation', SOAS, London.

Briscoe, J. (1983), 'Water supply and health in developing countries: selective primary health care revisited', *Proceedings of the International Conference on Oral Rehydration Therapy*, Washington, June 7-10 1983, pp. 141-150.

Briscoe, J. (1994), 'Implementing the new water resources policy consensus: lessons from good and bad practices', *IWRA Congress*, Cairo.

Bromley, D.W. (1986), 'Closing comments at the conference on common property resource management', *Proceedings of the conference on common property resource management*, National Academy Press, Washington, Chapter 25, pp.593-598.

Brook-Cowen, P. (1997), 'Enabling environment for a well-functioning water sector: institutional and regulatory framework', *GTZ/NWSA Water and Sanitation Sector Reform Workshop No 4*, Workshop Report, Conclusions and Recommendations, Sana'a, November 22-25 1997.

Brooks, D.B. (1995), 'Planning for a different future: soft water paths', *Joint Management of Shared Aquifers*, Final Report, Chap 3, pp.45-50.

de Bruin, J.A., ten Heuvelhof, E.F. (1995), *Netwerkmanagement*, Lemma Uitgeverij B.V., Utrecht.

Brumfiel, E.M. (1980), 'Specialization, market exchange and the Aztec state: A view from Huexolta', in *Current Anthropology*, Vol. 21, No 4, pp.459-738.

Bryant, R.L. (1991), 'Putting politics first: the political ecology of sustainable development', *Global Ecology and Biogeograpy Letters*, Vol. 1.

Bryant, R.L. (1992), 'Political ecology: An emerging research agenda in third-world studies', *Political Geography*, Vol. 11, No 1.

Bryant, R.L. (1993), 'Power, knowledge and political ecology in the third world: A review', in *Progressive Physical Geography*, Vol. XX (February 1993).

Bryant, R.L. (1998), 'Power, knowledge and political theory: A review', Dept of Geography, King's College, London.

Bryant, R.L., Bailey, S. (1997), *'Third world political ecology'*, Routledge, London, 237 pp.

Burrill, A. (1998), 'The components of the societal value of water', *Geoforum*, Middlesex University Workshop.

Caponera, D.A. (1973), 'Water laws in Moslem countries', Vol. 1, 2 vols, *Irrigation and Drainage Paper 20/1*, FAO, Rome, 223 pp.

C.E.S. (1997), Consulting Engineers Salzgitter, Seba Messtechnik, Jordanian Consulting Engineer, RAMCO, 1997, Rehabilitation of Taiz town distribution network, Programme for the reduction of water losses in selected NWSA towns.

Chambers, R. (1983), *Rural Development: Putting the Last First*, Longman, New York, 246 pp.

Chambers, R. (1992), 'Rural appraisal: rapid, relaxed and participatory', Discussion Paper, 311, Institute of Development Studies, Brighton, 85 pp.

Chambers, R., Carruthers, I. (1986), 'Rapid appraisal to improve canal irrigation performance: experience and options', *IIMI Research Paper No.3*, International Irrigation Management Institute, Sri Lanka, 18 pp.

Clammer, J. (1984), 'Approaches to ethnographic research', Chap 4, in R.F. Ellen (ed), *Ethnographic Research: A Guide to General Conduct.* (Series ed Ellen, R.F., Research Methods in Social Anthropology, 1), Academic Press, London, pp.63-85.

Coase, R.H. (1992), 'The institutional structure of production', *The American Economic Review*, Volume 82, Issue 4 (September 1992), pp.713-719.

Coopers and Lybrand (1992), 'Institutional framework for effective resource management in the Middle East and North Africa', (Draft Final Report), Coopers and Lybrand, London.

Coopers and Lybrand (1993), 'Institutional framework for effective water resource management in the Middle East and North Africa', World Bank Report, Appendix I.

Cornwallis, K., Hogarth, D.G., (1917), *Handbook of Yemen*, Cairo Government Press, Cairo, 167 pp.

Coward, E.W. (1986), 'Direct or indirect alternatives for irrigation investment and the creation of property', Chap 13 in Easter (ed), *Irrigation Investment, Technology and Management Strategies for Development*, Westview Press, Boulder and London, pp.225-244.

Cowen, T. (1994), 'Improving the performance of non-formal institutions: a research proposal for the study of urban water and sanitation', World Bank, 38 pp.

Crane, R. (1994), 'Water markets, market reform and the urban poor: results from Jakarta, Indonesia', *World Development* 22, No 1, pp.71-83.

Cromwell, G. (1990), 'Some techniques for rapid appraisal of artisanal infrastructures', *RRA Notes* 1 (9, August), pp.18-26.

CSO (1996a), Statistical yearbook 1996, Central Statistical Organisation, Ministry of Planning and Development, ROY.

CSO (1996b), Final results of the population, housing and establishments census December 1994, Central Statistical Organisation, Ministry of Planning and Development, ROY, General Report.

CSO (1997), Final report and results of the first industrial survey, 1996, Central Statistical Organisation, Ministry of Planning and Development, ROY, General Report.

Dar Al Handasah (1997), 'Sana'a water supply and sanitation services. Water consumption forecasts', Dar Al Handasah, Sana'a.

Dar El-Yemen (1997), 'Hydrogeological and land-use studies of the Ta'iz region (Upper Wadi Rasyan Catchment)', *UNDDSMS Project YEM/93/010*, Strengthening of Water Resources Management Capabilities, Vol. I Main Report, Vol. II Supporting Data.

Davies, D., Sahooly, A. (1996), 'Decentralisation and competition for water in Yemen', *2nd UNDP symposium on Water Sector Capacity Building*, Delft, 4-6 December, 1996.

Deason, J.P. (1991), 'Water policies relating to environmental and health issues', in J. Keenan (ed), *Technological Issues in Water Management*, World Bank Conference, Washington, pp.53-55.

Delft, I.H.E., UNDP (1991), 'A strategy for water sector capacity building', *Proceedings of the UNDP Symposium*, Delft, 3-5 June, 1991.

Dellapenna, J.W. (1995), 'Why are true markets in water rare? or Why should water be treated as public property?', *Joint Management of Shared Aquifers*, Final Report.

Department of Irrigation and Soil and Water Conservation (1993), 'Liquid Gold. A comparative research programme on irrigation water management in South Asia', Wageningen Agricultural University, The Netherlands, 16 pp.

DHV (1993), 'Runoff assessment in areas with scarce flow measurements', YEM/87/015 Final Report.

Directorate of Overseas Surveys (1981), *Series Y.A.R. 50. Edition 1-D.O.S.*, 1981.

Dixon, J.A., Hamilton, K. (1996), 'Expanding the measure of wealth', *Finance and Development*, December 1996.

Doorenbos, J., Pruitt, W.O. (1977), 'Crop water requirements', *Irrigation and Drainage Paper No 24*, FAO, Rome, 179 pp.

Douglas, M. (1978), 'Cultural Bias', *Occasional Papers of the Royal Anthropological Institute*, No.34, London.

Dresch, P. (1989), *Tribes, government, and history in Yemen*, Clarendon Press, Oxford, 440 pp.

Dubay, L. (1993), 'Mission report on Taiz water supply', NWSA, Sana'a.

Dubay, L. (1989), 'Taiz water supply resources', NWSA report, Background to well EX8 (PW27), exploration potential of Wadi Warazan, Dabaab and Bani Khawlan, NWSA, Sana'a.

Dubay, L. (1996), 'Review of pump testing. NWSA report on Taiz emergency drilling and Habiir pumping tests', NWSA, Sana'a.

Al-Dubby, S.A., Taher, M.S. (1998), 'Agricultural water demand in Taiz region (Upper Wadi Rasyan)', National Water Resources Authority, PPS Technical Note Series No. TN-98-05, Sana'a.

Easter, K.W. (1991), 'Intersectoral water allocation and pricing', in J. Keenan (ed), *Technological Issues in Water Management*, World Bank Conference, Washington, pp.9-14.

Eckersley, R. (1992), *Environmentalism and political theory: Toward an ecocentric approach*, UCL Press, London.

Eger, H. (1986), 'Runoff agriculture: A case study about the Yemeni Highlands', Ph.D. Dissertation, University of Tubingen.

Al-Eryani, M.L., Bamatraf, A., Al-Saqaf, G., Handash, S. (1995), 'Water rights aspects of additional water supply for the city of Sana'a', World Bank, Draft.

Falkenmark, M., Lundqvist, J. (1995), 'Looming water crisis: new approaches are inevitable', in L. Ohlsson (ed), *Hydropolitics*, University Press, Dhaka, pp.178-224.

Feitelson, E., Allan, J.A. (1998), 'Economic and political dimensions in changing perceptions of water in the Middle East', *Geoforum*, Middlesex University Workshop.

Feitelson, E., Haddad, M. (1998), 'Joint management of shared aquifers: A flexible, sequential building approach. A stepwise open-ended approach to the identification of joint management structures for shared aquifers', *Joint Management of Shared Aquifers*.

Foucault, M. (1971), 'Orders of discourse', *Social Science Information*, Vol. 10.2.

Fuchs, M., Hadas, A. (1972), 'The heat flux density in a non-homogeneous bare loessial soil', *Boundary-Layer Meteorology* 3, pp.191-200.

Giddens, A. (1976), *New rules of sociological method. A positive critique of interpretative sociologies*, Hutchinson, London.

Giddens, A. (1984), *The constitution of society. Outline of the theory of structuration*, Cambridge: Polity Press.

Giddens, A. (1990), *The consequencies of modernity*, Cambridge: Polity Press.

Gitec Dorsch, (1989), 'Improvement and expansion of water supply and sanitation in provincial towns in the Yemen Arab Republic', Volumes A-F, Gitec Dorsch.

Golman, D. (1997), *Vital lies, simple truths*, London, Bloomsbury.

Guggenheim, S. (1991), 'Institutional arrangements for water resources development', in J. Keenan (ed), *Technological Issues in Water Management*, World Bank Conference, Washington, pp.21-24.

Gun, J. van der, Abdul Aziz, A. (1995), 'The water resources of Yemen', Vol. Report WRAY-35, TNO, Delft, 108 pp.

Haddash, S. (1998), 'Water ownership rights', *Yemen Times*, February 23rd, 1998. 'Water diversion rights', *Yemen Times*, March 2nd, 1998.

Hajer, M.A. (1995), *The politics of environmental discourse: ecological modernization and the policy process*, Clarendon Press, Oxford.

Al-Hamdi, M.I. (1998), 'Qat: technical, economical and social issues', World Bank, Sana'a, p.14.

Handley, C.D. (1996a), 'Baseline socio-economic survey in Ta'iz governorate (excluding Tihama)', UNDDSMS Project YEM/93/010.

Handley, C.D. (1996b), 'Water use patterns in the Amran valley, Yemen', World Bank Mission July-August 1996, World Bank, Sana'a.

Handley, C.D. (1997a), 'Base-line study of Amran and Yarim public water supply institutions', Ministry of Electricity and Water, Republic of Yemen, Technical Secretariat for Water Supply and Sanitation Sector Reform.

Handley, C.D. (1997b), 'Water markets in Yemen: Ta'iz, San'a, Amran and Yarim. Reform initiatives and options in Yemen', *GTZ/NWSA Water and Sanitation Sector Reform Workshop No 4*, Workshop Report, Conclusions and Recommendations, Sana'a, November 22-25 1997.

Handley, C.D. (1999a), 'Household water use survey Ta'iz, Yemen', UNDP / Netherlands / UNDDSMS Project YEM/93/010.

Handley, C.D. (1999b), 'Review of technical and institutional options for Ta'iz city water supply', Taiz Water Supply Pilot Project, World Bank Report, Sana'a.

Handley, C.D. (1999c), 'Autopsy of an aquifer: a case study of Al Hayma, Yemen', *Proceedings of the IUGG conference*, Birmingham, 19-21 July 1999.

Handley, C.D., Dotteridge, J. (1997), 'Causes and consequences of extreme water shortage in Ta'iz, Yemen', in Chilton et al (eds), *Groundwater in the Urban Environment: Problems, Processes and Management*, pp. 325-330.

Hansen, T. (1983), *From Copenhagen to Sana'a*, Daar Al Awda, Beirut, 376 pp.

Hansma, K., Hermans, L. (1997), 'Stakeholder participation for water resources management in the Ta'iz region', M.Eng Thesis, Delft University of Technology, 86 pp.

Hardin (1968), 'The tragedy of the commons', *Science* 162, pp.1243-1248.

Haskoning (1990), 'Investigation of the environmental impact of industries in the Republic of Yemen', Annex: Site visits to main industries of Sana'a, Hodeida and Ta'iz, NWSA, Sana'a.

Heidbrink, K. (1994), 'The social and economic impact of qat in Yemen', PG Dip.Ag.Econ Thesis, Reading University, 43 pp.

Hildyard, N. (1998), 'The world bank and the state: a recipe for change?', Cornerhouse Public Outreach and Research Unit, Sturminster Newton, Dorset.

Jewitt, S. (1994), 'Development theories and practice', UCL/SOAS Seminar, 14/12/94.

Johnson, C. (1996), 'Institutions, social capital and the pursuit of sustainable livelihoods', Sustainable Livelihoods Programme, Institute of Development Studies, University of Sussex.

Joshi, A.R. (1995), 'Informal Support Systems in Yemen', World Bank, Sana'a.

Jurriens, R., de Jong, K. (1989), 'Irrigation water management, a literature survey', Working Party 'Irrigation and Development', Wageningen Agricultural University, Wageningen, pp.35-45.

Kalbermatten Associates, Inc. (1998), 'Formation of Regional Water and Sanitation Corporations', Technical Secretariat for Water Supply and Sanitation Sector Reform, Ministry of Electricity and Water, Republic of Yemen.

Karshenas, M. (1992), 'Environment, employment and sustainable development', (Technology and Employment Programme Working Paper 237), *International Labour Office*, Geneva, 57 pp.

Keenan, J. (1991), *Technological issues in water management*, (ed J. Keenan), World Bank Conference, Washington, pp.1-14.

Kennedy, D.N. (1991), 'Allocating California's water supplies during the current drought', in J. Keenan (ed), *Technological Issues in Water Management*, World Bank Conference, Washington, pp.15-18.

Kerkvliet, B.J.T. (1990), *Everyday politics in the Philippines. Class and status relations in a Central Luzon village*, University of California Press, Berkeley.

Kingdon, J. (1984), *Agendas, alternatives and public policies*, New York, Harper-Collins.

Laird, J. (1991), 'Environmental accounting: putting a value on natural resources', *Our Planet* 3(1), pp.16-18.

El-Lakany, M.A. (1978), 'Sorghum, maize and millet', Terminal Report of Breeder/Agronomist, FAO Project YEM/010, AREA, Ta'iz.

Lam, W.F. (1994), *Institutions, engineering infrastructure and performance in the governance and management of irrigation systems: the case of Nepal*, Unpublished Ph.D. thesis, School of Public and Environmental Affairs and Dept. of Political Science, Indiana University, USA.

Leggette, Brashears, Graham (1977), 'Hydrogeologic investigation for well field development in Al Hayma basin', NWSA, Sana'a.

Leggette, Brashears, Graham (1980), 'Completion of As Sahlah wellfield, Taiz, YAR, NWSA, Sana'a.

Leggette, Brashears, Graham (1981), 'Well field development in Al Haima basin', Summary Report *NWSA, Sana'a.*

Leung, K. (1999), 'Monitoring qat with earth observation data and geographic information system techniques in the region of Jabal Sabir, Ta'iz, the Republic of Yemen', SOAS MSc thesis.

Lichtenthaler, G. (1999), 'Water management and community participation in the Sa'dah basin of Yemen', World Bank, San'a.

Littlefair, K. (1998), 'Water use and water pricing in southern Kerala', London: Department of Geography, *SOAS*, University of London. Unpublished dissertation.

Lundqvist, J. (1998), 'The triple squeeze on water: rain water, provided water and waste water in socio-economic and environmental systems', *Geoforum*, Middlesex University Workshop.

Mackintosh, M. (1992), 'Introduction', in M.Wuyts, M. Mackintosh, T. Hewiit, (eds), *Development, policy and public action*, Oxford University Press, in association with the Open University, Oxford, pp.1-9.

MacMillan, A.A. (1976), 'Irrigation projects identification mission, Yemen Arab Republic', (Report No.22/76 YAR-3) FAO, Rome.

Mahdi, M. (1986), 'Private rights and collective management of water in a High Atlas Berber tribe', Chapter 10, pp.181-198, *Proceedings of the conference on common property resource management*, National Academy Press, Washington.

Maktari, A.M.A. (1971), *Water rights and irrigation practices in Lahj*, Cambridge University Press, Cambridge, 202 pp.

Mallat, C. (1995), 'The quest for water use principles: Shari'a and custom in the Middle East', in J.A. Allan, and C. Mallat (eds), *Water in the Middle East: legal, political and commercial implication*, London: Tauris Academic Studies, pp.127-150.

Manzungu, E. (1999), *Strategies of smallholder irrigation management in Zimbabwe*, Published doctoral thesis, Wageningen University, 202 pp.

McLachlan, K. (1988), *The neglected garden: the politics and ecology of agriculture in Iran*, I.B. Tauris, London. 303 pp.

Mclagan, I. (1994), Personal communication.

McSweeney, J. (1998), 'Social and environmental use-values of water: reflections on the disenchantment of water and the environment', *Geoforum*, Middlesex University Workshop.

Merrett, S. (1997), *Introduction to the economics of water resources*, UCL Press, London, 211 pp.

Migdal, J.S. (1988), *Strong societies and weak states: state-society relations and state capabilities in the third world*, Princeton University Press, Princeton.

Miller (1979), 'Self-help and popular participation in rural water systems', Development Centre Studies (Paris: OECD).

Milroy, T. (1994), The hanging gardens of Arabia, Video, Arid Lands Initiative.

Mitchell, B., Escher, H. (1978), 'A baseline socio-economic survey of the Taiz-Turbah road influence area', Vol. II, Appendix, World Bank.

Mockus, V. (1972), 'Estimation of direct runoff from storm rainfall', *National Engineering Handbook*, Section 4, Hydrology. (ed: U.S. Department of Agriculture), National Technical Information Service, Washington.

Moench, M. (1997), 'Local water management: options and opportunities in Yemen', Summary Report to World Bank Decentralized Management Study, World Bank, Sana'a.

Le Moigne, G., Subramanian, A., Xie, M., Giltner, S. (1994), *A guide to the formulation of water resources strategy*, World Bank, Washington, 102 pp.

Mollinga, P.P. (1998), *On the waterfront: water distribution, technology and agrarian change in a South Indian canal irrigation system*, Published doctoral thesis, Wageningen University, 307 pp.

Monteith, J.L. (1965), 'Evaporation and environment. State and movement of water in living organisms', *19th Symposium of the Society for Experimental Biology*, Cambridge University Press, pp.205-234.

Monteith, J.L. (1991), 'Weather and water in the Sudano-Sahelian Zone', in M.V.K Sivakumar, J.S. Wallace, C. Renard, C. Giroux, *Soil Water Balance in the Sudano-sahelian Zone*, IAHS, Wallingford, pp.11-29.

Monteith, J.L., Unsworth, M.H. (1990), *Principles of environmental physics*, 2nd ed. Edward Arnold, London, 291 pp.

Montgomery, James M. (1975), 'Water supply and sewerage facilities for Taiz YAR', Feasibility Study, Vol. II Water Resources Evaluation, USAID. Contract No. AID/ASIA-C-1081.

Morris, T. (1991), *The despairing developer. Diary of an aid worker in the Middle East*, I.B. Tauris, New York, 286 pp.

Morton (1994), *The poverty of nations: The aid dilemma at the heart of Africa*, British Academic Press, London, 265 pp.

Mu, X., Whittington, D., Briscoe, J. (1990), 'Modelling village water demand behaviour: a discrete choice approach', *Wat. Resour. Res.*, 26, pp.521-529.

Mullick, M.A. (1987), *Socio-economic aspects of rural water supply and sanitation. A case study of the Yemen Arab Republic*, The Book Guild, Lewes, Sussex, 252 pp.

Mundy, M. (1989), 'Irrigation and society in a Yemeni valley', *Peuples Mediterraneens* 46, pp.97-128.

Mundy, M. (1995), *Domestic government: kinship, community and polity in North Yemen*, I.B. Taurus, London, 317 pp.

Myrdal, G. (1968), *Asian drama: an inquiry into the poverty of nations*, Vol. 2 (New York: Twentieth Century Fund), 896 pp.

Naff, T. (1991), 'The Jordan basin: political, economic and institutional issues', in J. Keenan (ed), *Technological Issues in Water Management*, World Bank Conference, Washington, pp.115-118.

Noman, S.A. (1982), 'Review of hydrogeological studies conducted in the Sana'a basin Yemen Arab Republic', ACSAD II (2, *Deuxieme Symposium Arabe sur les Ressources en Eau, Rabat*, 17-21 Septembre 1981), pp.210-231.

NWRA (1996), 'Well inventory Upper Wadi Rasyan', Ta'iz, National Water Resources Authority, Republic of Yemen.

NWRA (1998a), 'Water resources management issues in the Ta'iz region (Upper Wadi Rasyan Catchment)', Draft. National Water Resources Authority, Republic of Yemen.

NWRA (1998b), 'Water resources management strategies in the Ta'iz region (Upper Wadi Rasyan)', Draft. National Water Resources Authority, Republic of Yemen.

NWRA (1999a), 'Agriculture and irrigation in the Ta'iz region (Upper Wadi Rasyan)', Draft. National Water Resources Authority, Republic of Yemen.

NWRA, (1999b), 'Socio-economic study of the Ta'iz region (Upper Wadi Rasyan)', National Water Resources Authority, Republic of Yemen.

Ohlsson, L. (1995), 'The role of water and the origins of conflict', in L. Ohlsson (ed), *Hydropolitics*, University Press, Dhaka, pp.1-28.

Ohlsson, L. (1998), 'Water and social resource scarcity – an issue paper commissioned by FAO/AGLW', Presented as a discussion paper for the 2nd FAO E-mail conference on managing water scarcity, WATSCAR2.

Ohlsson, L. (1999), 'Environment, scarcity and conflict: a study of malthusian concerns', Department of Peace and Development Research, Goteborg University.

Oorthuizen, J., Wester, P. (1994), 'Irrigation and development: reader for K200-310', Department of Irrigation and Soil & Water Conservation, Wageningen Agricultural University, The Netherlands.

Ostrom, E. (1986), 'Issues of definition and theory: some conclusions and hypotheses', Chapter 26, pp.599-615. Proceedings of the conference on common property resource management, National Academy Press, Washington.

Ostrom, E. (1999), 'Social capital: a fad or a fundamental concept?', Workshop in Political Theory and Policy analysis, in P. Dasgupta, I. Serageldin (eds), *Social capital: a multifaceted perspective*, Washington DC World Bank, pp.172-214.

Otchet, A. (1999), 'An economic mirage?', *The UNESCO Courier*, February 1999, pp.31-32.

Patai, R. (1976), *The Arab mind*, Scribners, New York, 376 pp.

Pearce, D. (1993), *Economic values in the natural world*, Earthscan, London, 129 pp.

Pretty, J., Mearns, R., McCracken, J., Scoones, I. (1989), Notes on the discussions held by the 35 participants at the joint IDS/IIED RRA review workshop, 19-20 June 1989, *RRA Notes* 1 (7 September, Proceedings of RRA Review Workshop), pp.54-62.

Prinz, D. (1994), 'Water harvesting – past and future', in L.S. Pereira (ed), *Sustainability of Irrigated Agriculture. Proceedings*, NATO Advanced Research Workshop, Vimeiro, 21-26 March 1994, Balkema, Rotterdam, pp.1-31.

Radwan, L. (1994), 'Water use efficiency in delta Egypt: irrigation and social control', *Erasmus Workshop*, Padova, Italy.

Rajeswary, I. (1992), 'Yemeni farmers learn to conserve precious water', *SOURCE* 4 (2 December 1992), pp.17-19.

Rango, A. (1984), 'Runoff synthesis utilizing landsat hydrologic land use data and soil conservation service (SCS) models', United States Soil Conservation Service, 7 pp.

Redclift, M. (1987), *Sustainable development: exploring the contradictions*, Methuen and Co, London.

Reisner, M. (1986), *Cadillac desert: the American West and its disappearing water*, New York: Penguin Books. First published 1986 revised and up-dated 1993.

Rhebergen, G.J., Waveren, E.J. van (1990), 'Agroclimatic characterisation of the western part of the Yemen Republic', YEM/87/007 Field document No 4. Dammar, Yemen: Soil Survey and Land Classification Project, FAO.

Rogers, P. (1992), 'Integrated urban water resources management', Keynote Paper 7, *International Conference on Water and the Environment: Development Issues for the 21st century*, 26-31 January, Dublin, Ireland, 39 pp.

Rondinelli (1983), *Development projects as policy experiments: An adaptive approach to development administration*, London, Methuen.

Saqqaf, A.Y. (1985), 'Fiscal and budgetary policies in the Yemen Arab Republic', in B.R. Pridham (ed), *Economy, Society and Culture in Contemporary Yemen,* Croom Helm, Dover, New Hampshire, pp.83-95.

SAWAS, (1997), 'Sources for Sana'a water supply', Final Technical Report. NWSA. TNO, Institute of Applied Geoscience, Delft.

Al-Sawwaf, M.M. (1977), *The Muslim book of prayer*, M.M.Al-Sawwaf, Mecca. 70 pp.

Sax, J.I. (1994), 'Understanding transfers: community rights in the privatisation of water', *West-Northwest*, Vol. 1, p.13.

As Sayagh, A.K. (1998), 'Industrial water requirement for Taiz region (Upper Wadi Rasyan)', National Water Resources Authority, PPS Technical Note Series No. TN-98-02, NWRA, Sana'a.

Serageldin, I. (1994), *Water supply, sanitation and environmental sustainability: the financing challenge*, World Bank, Washington, 100 pp.

Serjeant, R.B. (1964), 'Some irrigation systems in Hadramawt', *Bulletin of the School of Oriental and African Studies* 27, pp.33-76.

Shuttleworth, W.J. (1990), 'Evaporation and evapotranspiration', in D.R. Maidment (ed), *Handbook of Hydrology*, McGraw-Hill, New York.

Shuttleworth, W.J., Wallace,J.S. (1985), 'Evaporation from sparse crops – an energy combination theory', *Quart.J.Roy.Met.Soc*, 111, pp.839-855.

Smith, M. (1992), *CROPWAT A computer program for irrigation planning and management*, FAO, Rome, 125 pp.

Steenbergen, F. van (1996), *Institutional change in local water resource management: cases from Balochistan*, Nederlandse Geografische Studies.

Swanson, J.C. (1985), 'Emigrant remittances and local development: co-operatives in the Yemen Arab Republic', in B.R. Pridham (ed), *Economy, Society and Culture in Contemporary Yemen*, Croom Helm, Dover, New Hampshire, pp.132-146.

Swinscow, T.D.V. (1983), *Statistics at square one*, British Medical Association, London, 86 pp.

Taher, M.S. (1998), 'Domestic water demands and wastewater loads in Taiz region (Upper Wadi Rasyan)', National Water Resources Authority, PPS Technical Note Series No. TN-98-03, NWRA, Sana'a.

TESCO-Vitziterv-Vituki (1971), 'Survey of the agricultural potential of Wadi Zabid, Yemen Arab Republic', (Final Report and Technical Reports 1-12), Food and Agriculture Organisation of the United Nations; United Nations Development Programme (1971): TESCO-Vitziterv-Vituki, Budapest.

Thanh, N.C., Biswas, A.K. (1990), 'Water systems and the environment', in N.C. Thanh, A.K. Biswas (eds), *Environmentally-sound Water Management*, Oxford University Press, Delhi, pp.1-29.

Therkildsen, O. (1988), *Watering white elephants? Lessons from donor funded planning and implementation of rural water supplies in Tanzania*, Scandanavian Institute of African Studies, Uppsala, 224 pp.

Thompson, M. (1988), 'Socially viable ideas of nature: a cultural hypothesis', in E. Baark, U. Svedin (eds), *Man, nature and technology: essays on the role of technological perceptions*, Chapter 4, Macmillan, Basingstoke.

Thompson, M. (1995), 'Policy-making in the face of uncertainty: the Himalayas as unknowns', in G.P. Chapman, M. Thompson (eds), *Water and the quest for sustainable development in the Ganges valley*, Chapter 2, Mansell Publishing, London.

Tipton, Kalmbach (1979), 'Taiz water and sewerage project irrigation and compensation study', Hazen and Sawyer Report, NWSA, Sana'a.

Turton, A.R. (1999a), 'Precipitation, people, pipelines and power: towards a political ecology discourse of water in Southern Africa', *MEWREW Occasional Paper* 11.

Turton, A.R. (1999b), 'Water state and sovereignty: the hydropolitical challenges for states in arid regions', *MEWREW Occasional Paper* 5.

Turton, A.R. (1999c), 'Water scarcity and social adaptive capacity: towards an understanding of the social dynamics of managing water scarcity in developing countries', *MEWREW Occasional Paper* 9.

Turton, A.R. (1999d), 'Water demand management: a case study from South Africa', *MEWREW Occasional Paper* 4.

Uphoff, N. (1981), 'The institutional-organiser (IO) program in the field after three months: a report on trip to Ampare/Gal Oya June 17-20, 1981', Ithaca: rural development committee, Cornell University, Mimeo.

Uphoff, N. (1986), 'Getting the process right: improving irrigation water management with farmer organization and participation', Working Paper, Cornell University.

Uphoff, N., Wickramasinghe, M.L., Wijayaratna, C.M. (1990), "Optimum' participation in irrigation management: issues and evidence from Sri Lanka', *Human Organisation*, Vol. 49, No.1, pp.26-40.

USAID (1983), Yemen general soil map, *USAID* / YAR-MOA. Contract #279-0042.

Varisco, D.M. (1982), 'The adaptive dynamics of water-allocation in Al-Ahjur, Yemen Arab Republic', Ph.D. Thesis, University of Pennsylvania.

Varisco, D.M. (1983), 'Sayl and Ghayl: The ecology of water allocation in Yemen', *Human Ecology* 11 (4), pp.365-383.

Varisco, D.M. (1991), 'The future of terrace farming in Yemen: A development dilemma', *Agriculture and Human Values* 8, pp.166-172.

Vincent, L. (1990), 'The politics of water scarcity: irrigation and water supply in the mountains of the Yemen Republic', *(ODI/IIMI Irrigation Management Network Paper* 90/e) ODI/IIMI, London. 28 pp.

Vincent, L. (1991), 'Debating the water decade: The Yemen Republic', *Development* 9 (2 June), pp.197-216.

Wallace, J.S. (1991), 'The measurement and modelling of evaporation from semiarid land', in M.V.K. Sivakumar, J.S. Wallace, C. Renard, C. Giroux (eds), *Soil Water Balance in the Sudano-sahelian Zone*, IAHS, Wallingford, pp.131-148.

WAPCOS (1996), 'Dam site selection', Southern Uplands Regional Development Programme, Ta'iz Yemen.

Ward, C. (1998), 'Practical responses to extreme groundwater overdraft in Yemen', *Yemen: the challenge of social, economic and democratic development.* University of Exeter, Center for Arab Gulf Studies, Conference Proceedings, April 1-4, 1998.

Ward, C., Abdul Bari, M., Schlund, M., Othman, M. (1998), 'Yemen: agricultural strategy Working Paper Number 4', Rural Finance, World Bank, San'a.

Ward, C., Moench, M. (In Preparation): 'Results of decentralised management study', World Bank, San'a.

Water Care Associates (1995), 'Waste water survey for Hayel Said Anam', Hayel Said Anam.

Weber, M. (1978), *Economy and society: an outline of interpretive sociology*, in G. Roth and C. Wittich (eds), University of California Press, Berkeley.

Weir, S. (1985), 'Economic aspects of the qat industry in North-West Yemen', in B.R. Pridham (ed), *Economy, Society and Culture in Contemporary Yemen*, Croom Helm, Dover, New Hampshire, pp.64-82.

Welle, J. van der, (1997), 'Hydrochemistry and pollution studies in Upper Wadi Rasyan Catchment', Unpublished Report, National Water Resources Authority, San'a.

Wilkinson, J.C. (1977), *Water and tribal settlement in South-East Arabia. A study of the aflaj of Oman*, Clarendon Press, Oxford, 276 pp.

Williams, J.B. (1979), 'Yemen Arab Republic - Montane plains and Wadi Rima project: a land and water resources survey. Climate of the Montane Plains and Wadi Rima', (Project Record 42. YAR-01-48/REC-42/79), Land Resources Development Centre, Tolworth, 116 pp.

Wilson, E.M. (1990), *Engineering hydrology*, 4th ed. Macmillan, Basingstoke, 348 pp.

Winpenny, J. (1994), *Managing water as an economic resource*, Routledge, London. 133 pp.

World Bank (1986), 'Southern agricultural regional development project', FAO / World Bank, Annex I.

World Bank (1993), 'Republic of Yemen agriculture sector study: strategy for sustainable production', Volume I Main Report, Volume II Annexes and Statistical Data, World Bank.

World Bank (1997), 'Yemen: towards a water strategy: an agenda for action', Report No 15718-YEM, World Bank, Washington.

World Bank (1998a), 'Republic of Yemen: Agricultural Strategy Note', Report No 17973-YEM. First Draft, World Bank, San'a.

World Bank (1998b), 'Taiz pilot water supply project', Mid term review mission report. World Bank, San'a.

World Wildlife Fund (1996), 'Dangerous curves: does the environment improve with economic growth?', WWF Discussion Paper.

Yoshida, H. (1999), 'Information flows for African communities and their environments. An analysis of roles of geographical information processes of decision making', Unpublished Ph.D. thesis of the University of London.

Zagni, A.F.E. (1996), 'Yemen: decentralised management for water strategy: water resources component', Consultants Report for the World Bank, World Bank, San'a.

# Index

247